ST. PATRICK, THE GREEN REVOLUTION, AND THE HYDROGEN CONVERSION PROJECT

The Green Everevolution

Everevolution is a neologism combining the elements of the words evergreen, evolution and revolution. St. Patrick, the Green Revolution and the hydrogen conversion project crnate a Green Everevolution. The word, conversion, was chosen because it has religious, biological (life/green) and chemical connotations.

Cycles in Nature

"In New Hampshire they drove the skunks out. What they didn't seem to know is that the skunks ate turtle eggs. With the skunks gone, the turtles multiplied and ate the fish. So they brought the skunks back." Father Hilary Freeman, O.P., St. Thomas Aquinas Priory

St. Patrick Patron Saint of Ecology: The Cyclical Saint
Major Theses

There are cycles in nature, in history and in the history of ecology. St. Patrick brings these cycles into syncronicity. <u>The patron saint of the ecology movement is St. Patrick who rose from slavery and persecution to the spiritual definition of the Irish people</u>. Likewise, the whole focus of the ecology movement should go from <u>chained</u> relations to willed bonded relations and to the interrelated of all species in the bonded relations of nature. St. Patrick favored purity of intent.[1] In parallel, in the chemical world the use of hydrogen is emphasized because of its purity. Hydrogen has purity of ecological intent. St. Patrick also opposed the excessive use of military force.[2] A psychic dream prompted his mission to deliver the Irish people and all people from bondage.[3] The ecology movement should work to deliver all people and other species "from human bondage."[4] This is the theme of this book and a theme in the life of the writer, Patrick O'Dougherty, to rise up from all forms of personal and collective bondage and to deliver plant and animals life forms from human bondage. This is the ecology movement's psychic dream which is like Martin Luther King's dream to be "free at last." Weighing and periodicity are key elements in nature, hence, St. Patrick and the cyclical weighing of the hydrogen conversion project.

Richard Feynman and St. Patrick

Richard Feynman's nanotechnology is the attempt to rearrange the atomic structure of all materials, to develop new forms, methods and beings.[5] Richard

Feynman and genetic engineers and nanotechnologizers want to "reassemble the Earth" gene by gene and molecule by molecule. [6] In addition, we should try to make "the inhabitable places habitable. Science fictions writers call this terra forma. " [7] **A**

1

flashback to the original structure of the earth at the time of Adam and Eve's fall is virgin technology. Will tampering with the universe is will technology. Martin Luther thought himself imprisoned in his body and hoped death would allow him to escape it. [8] In contrast, St. Patrick rose up from slavery to bonded relationships in all their forms and to "virgin time" and willed relationships.[9] He made Ireland a habitable place. He reassembled the personal universe through the priesthood and through the freeing of his body from slavery. He was a Christian personalist which is a viewpoint that will renew ecology. The intuitionist school of mathematics will add depth to this perspective.

Cardinality

A cardinality is "defining by abstraction, one my say that which is equivalent to all equivalent sets." [10] For example, a cardinality is a cardinal number associated with a given class; since two sets are defined to have the same cardinality if their numbers can be put in one to one correspondence, this is an equivalent relation. And the cardinality of any given finite class is taken to be the largest number of the initial sequence of natural numbers beginning with on that so corresponds to it. The cardinality of the set knife, fork, spoon I: knife, fork, spoon I: equals I: 1,2,3, I: equals three.[11] What is the principle of cardinality? This is "Cantor's theorem that allows one to construct a hierarchy of cardinalities."[12] The reason cardinality is introduced here is because it relates to infinity, eternity and infinite sets. Ecology is essentially theology based in part on these ideas. Cardinality reflects the Catholic cardinals who are working towards the eternal and infinite.

Impossibility Theorem

Father Hilary Freeman argues there is an impossibility theorem in nature. One cannot build up all the populations of the earth up to the standards of living of Western Civilization. Another reference on this impossibility theorem follows: Kenneth Joseph Arrow, economist, has a revolutionary welfare theorem on this topic. He proves mathematically that a perfect economic government can never be possible. [13] A perfect ecological system is also not possible.

The Catholic Context

How would you define impossibility theorem, cardinality, hierarchy, freedom, freeman, and equivalence within a Catholic context? First, ecological perfection is an impossibility. Second, the priest or cardinal is the basis of evolution which is theocratic and largely aristocratic. Third, nature is largely a theocratic/aristocratic rather than a democratic process. Fourth, the Trinity does not vote on the salvation of souls. Thus nature is not largely democratic. Fifth, are all sets in nature, for example, species, equivalent? The equivalence of all sets in nature is an impossibility theorem because life is the more valued set. Sixth, what is an equivalent set in nature,

2

in species? There is very little democracy in nature. So how can one say sets of birds have internal equivalence. There is often a lead bird in a flock of birds. Thus sets in nature are not always cleanly defined. How do you define these concepts and nature from a Catholic perspective? The Christian Trinity and the dogmas of the Church define them. For example, nature is a personal activity which has an eternal social dimension which reflects the cardinality of infinity and infinite sets. Perfect human nature is impossible because of original sin. The Church is a hierarchy. Evolution is not completely materialistic so it is not deterministic. Freedom is thus the basis.

The Deep Ecology/Shallow Ecology Dualism

Deep Ecology is a movement typified philosophically by a paradigm shift from anthropocentrism to ecocentrism and environmental activism. Perception, value and lifestyle changes in the movement try to redirect the ecologically destructive, complex paths of modern industrial growth.[14] Arne Naess, philosophy leader at the University of Oslo, contrasts, Deep Ecology with Shallow Ecology which is anthropocentric, favors technocratic environmentalism and emphasizes pollution, resource depletion and health and affluence of the population of developed countries. In contrast, Vice President Al Gore sees the Deep Ecology as misanthropic which sees humans as "aliens" on the earth. Gore favors the human-dominant Christian stewardship philosophy. Deep Ecology favors ecocentrism .[15] Thus humans are "strangers in a strange land" which is a Biblical reference from Moses. [16]

Christian Stewardship

What is Christian stewardship? Christ is the third person of the Trinity. The Trinity has an eternal social dimension expressed as three persons in one God .[17] Who is a steward? A steward: is "one who manages another's property or financial affairs; one who administers anything as the agent of another or others." Stewardship: "to act or serve as steward. "[18] Christian stewardship is a principle of dominion over nature, for example, over starvation.

Deep Ecology: From Science to Wisdom

Arne Naess claims the universe which he identifies with is the "greater I am." He claims Deep Ecology relates to spohy or the Greek word, sophia, meaning wisdom which connects to ethics, norms and practice. Science does not try to answer why questions. Deep Ecology is in shift from science to wisdom.[19]

Resolving Dualisms

Are dualisms and contradictions in math, ecology, and philosophy resolved in God, in Christ? The answer is a question mark. Moreover, there are several dualisms

3

in philosophy: m.an and nature, mind/body, good/evil, matter/spirit, slave/free. There is a dualism already in ecology between Shallow and Deep Ecology.

Mark McGee: Dualistic Thinking is Fallacious

What mathematical or philosophical ideas would resolve these dualisms? Mark McGee maintains typological thinking, for example dualistic thinking, is fallacious. There is no completely ideal type in nature. The basis of nature is diversity. Biology is based on individual differences and variation. [20] Roger Barker thinks behavior depends on the setting and on the situation. There is a great variation in setting and situation For example, people behave differently at picnics than at weddin gs. [21] Also, there is no ideal model in society. So Karl Marx's typological, dialectical, class struggle model is fallacious. Additionally, the Deep Ecology and Shallow Ecology dualism is also fallacious and not based on the individuality and variation of nature.

Seeds

The seeds of the Deep Ecology movement appear in the ecocentrism and social activism of St. Francis of Assisi, Henry David Thoreau, St. Patrick, John Muir, D. H. Lawrence, Aldous Huxley and Theodore Rosazk. The most important, seminal inspiration for the movement in the 1960s came from Leopold's "land ethic" and Rachel Carson's, Silent Spring in 1962. Carson's book is an indictment of the invisible and widespread pesticide use. Her empathy went to the biological integrity of the earth and all of its species. Pesticides cause birth defects. She challenged humanity's "right" to control nature and dominate the Earth. She challenged Western Culture's anthropocentrism.[22] Actions have consequences, for example, ethical and social results for land.[23]

To Desacralize Nature

Lynn White, Jr. in 1966 maintained Christianity desacralized nature, abetted its exploitive world and anthropocentric world outlook. He thinks Christianity is build on a dualism between man and nature. White further claimed Marxism and other "post-Christian" ideologies in Western Civilization are essentially Judea-Christian heresies used to dominate and manipulate nature. To reform Christianity White argued for a cyclical revision to the outlook of St. Francis of Assisi who proselytized for "the equality of all creatures. "[24] Do creatures other than the human view nature as sacraments? The answer is they do not.

"Human Survival Environmentalism"

Another 1960s version of environmentalism is "human survival environmentalism" which focuses on pollution and human survival. It is anthropocentric. Barry Commoner is one of its exponents.[25]

4

What is the nature of man? It is quite complex. The Church teaches that man is a mammal with an <u>immortal soul.</u> He is a human animal. There is a break in evolution which created the distinctly human species.

Conservation Problems

What are the major conservation problems? E.0. Wilson, the Harvard conservationist and biologist maintains humans cause species extinction of nearly 10,000 species per year. Wild tigers and pandas appear "doomed." Global warming research and sound waves endanger marine mammals. Forest stands and sea food supplies are being rapidly depleted. Paul Erlich argued in 1985 that the U.S. has no excuse for developing one extra inch of underdeveloped land. [26]

"Dysfunctional Civilization"

Gore claims the U.S. is a "dysfunctional civilization" because of our over consumption and other problems. The term "dysfunctional civilization" is simplistic labelling. For example, is sustainable consumption possible considering the impossibility theorem? It may not be possible. Is sustainable agriculture going to lead to over consumption and species deletion in American and in the Third World? Many environmentalists think so. Is it a panacea? <u>In sustainable agriculture there is an inconsistency between environmental protection and sustained economic growth.</u> Moreover, an Ecology Council in the U.N. is a <u>chance.</u> The United Nations ecocentric World Charter for Nature adopted in 1982 lacks a <u>consistent</u> philosophical approach to ecology problems, environmental protection, population problems and an integrated Biosphere approach. [27] There are many <u>contradictions</u> in American civilization. However, the phrase "dysfunctional civilization" is too simplistic. The impossibility theorem fits many but not all of the problems in America. One can make a hose that fits a particular a watering system, for example.

Consumerism

Consumerism in industrial societies is a way of life and an addiction. Many animals are strangers in a strange consumerland. Industrial Growth Societies push for endless economic growth and progress. Modern visions promote megatechnology solutions such as high tech global management schemes for biosphere development. The conflict between the voluntary entrepreneur and the ecologist is the central earth issue of the twentieth century and of consumerism claims Thomas Berry. Anne and Paul Erlich argue the best approach to the ecology crisis, especially consumerism, is to "reduce the scale of human enterprise. "[28] In contrast, the writer has a "pro-life" approach to all living species. Consumerism should not delete the various life forms.

Three Major Thinkers

5

Who are several of the major thinkers in ecology? There are three critical thinkers in ecology. Thomas Berry, first, raises the contrast of wonderworld versus wasteworld. "Wonderworld" creates a wasteworld. Consumerism is creating a "spiritual and psychic-degradation." Fritjof Capra, second, defines Deep Ecology as organic integrated and holistic. Arne Naess, third, thinks the essence of Deep Ecology is questions.[29]

"Ecopsychology"

"Ecopsychology" is concerned with the psychological/spiritual dimensions of humanity's relationship to nature. The "ecological self" contains self-realization. Theodore Roszak claims the ecological unconscious is the core of the mind. Repression of this unconscious is understood frequently as madness. [30]

Richard Finn, in contrast, thinks another basis of the mind is elect ricit y.[31] There is a dualism in electricity, positive and negative currents. How would. one resolve this dualism in the brain? It is difficult to resolve. Is it a typology or a law of physics? The answer is it is a law of physics. Does it produce madness? The answer to this question is sometimes yes, for example, when epilepsy comes with madness.

In ecopsychology, separation from nature is an "original trauma," claims Chellis Glendenning. Consumerism is "techno-addiction." Ecopsychology renews the sense of connectedness to nature. [32] We must cultivate the wild creature within us to avoid becoming domesticated. For example, tribal ritual connects us to our whole being. It is a larger picture. Environmental activism is spiritual. Naess defines ecosophy which includes ecopsychology to mean "total view" and nonv iolence. [33] Andrew McLaughlin sees Deep Ecology as a social/political movement. It has a special logic. Deep Ecology has and eight point platform which is a number symbolizing infinity in ecology and in ecopsychology.

A Viable Community

Thomas Berry maintains the human community to be viable should move from the current anthropocentric norm to a geocentric norm. The sun is the main source of the earth's power. This is a heliocentric or sun centered norm. The universal galactic or star system and its creation fifteen million years ago is a mystery. The universe is an enduring reality and value. To need community is special because of the qualitative and quantitative human differences between humanity and the other species in the planetary system and in the galaxy .[34] The geocentric norm should integrate the human community rather than fragment it. However, the writer favors a universe or universal norm focusing on the dominant, most efficient, and very pure fuel, hydrogen. Hydrogen based viable communities are the norm of the future.

6

Hydrogen is the basis of the sun and the universe and composes ninety percent of the universe.

Plato

Humans have a special capacity for insight, speech, aesthetics, emotions and moral judgement. The cultural definitions of the human come from the surrounding enviro nment .[35] Plato emphasizes a universal ideal cultural approach. Cultures idealize. The ecological environment surrounding "Plato's cave" idealize. So should the insights, speech, aesthetics, emotions and moral judgements of the humans which come from the environment idealize.

The Ecologist Versus the Industrial Entrepreneur

Berry sees cultural pathology as the industrial plundering which is taking place in Western society. The viability of the human species and the earth connect. The extinction of life is not imminent. The ecologist stands versus the often pathological industrial entrepreneur.[36]

What has the industrial entrepreneur and industrialization done to Western Civilization? The industrial entrepreneur and industrialization have created a psychic and spiritual degradation. First, the sense of the social is dominant. Second, "progress" has desolated the world. Third, the universe is gendered. Fourth, the forms of tyranny are male dominating women, oppressors dominating nonoppressors, and humans dominating nature. Fifth, instead of emphasizing productivity with the world, the main relationship should be lawgiver and healer. Sixth, the earth is integral. Seventh, the medium is within the context of society and the context of the eart h .[37] Eighth, the earth has a shadow, evil, which ironically spelled backwards means live. Ninth, Greenpeace, Earth First, and People for the Ethical Treatment of Animals are confrontational groups to the present destructive activities of the earth, like redwood forest destruction. [38]

Mother Symbol

The universe is a Great Mother. The tree of life symbolizes the organic unity of mother earth and the universe. Danger to the tree will affect the whole organism. Primordial archetypal symbols, like the mother, in the human unconscious work as controlling factors in human thought, emotion and decision makin g. [39] The earth is the mother seed to the universe.

New Paradigm

Deep Ecology is a new paradigm maintains Fritzjof Capra. A paradigm shift is occurring from a mechanical system of building blocks, the body as machine metaphor, the view of compet1t1on as a model of society, and progress through technological and economic growth to the Deep Ecology position which claims to be similar to a "seamless garment," like Christ's cloak, which came out of nature. [40] However, where is Christ in Deep Ecology?

Cybernetics

Cybernetics or the theory and analysis of living systems is the best example of the new ecological paradigm. [41] The Gaia Hypothesis is an expression of cybernetics. It is the idea that the earth is a living organism within itself. [42] This organism has unity and integrity in this world and in the next. The living and the dead come together in the communion of the faithful.

Values Shift

Our values have shifted from expansion to conservation, quantity to quality, competitive to cooperative, domination to noncompetition and nonv iolence. [43] Green is the color of the life force. It is a value. Grey is the color of the mind another value and force. God is often a grey reality. Science is a grey reality. "Collective soul" is a concept which focuses values and a paradigm values shift like that found in the shift to ecocent rism. [44] It is a concept which symbolizes the coalesce of many movements like the peace movement, the feminist movement, the holistic health movement, the human potential movements, spiritual movements, ethnic and Third World movem ent s. [45] Collective soul blends colors and values and movements in a tapestry of relationships and connectedness.

Democracy and Aristocracy

Deep Ecology maintains there is a core democracy within the biosphere. This is true for some societies. but is it true for nature ?[46] Nature is largely aristocratic. For example, most animals do not have the democratic social dimensions and social values that the human animal does.

Recovery

Chellis Glendenning thinks she is in recovery from Western Civilization. By the systematic removing of our lives from the natural cycles she finds in life and in nature we are in a "Original Trauma." The American government and culture is by most social and psychological indicators distressed or insane. The symptoms of post-traumatic stress disorder are anger, psychic numbing and reactions like the loss of the sense of belonging to the univ erse. [47] In the psychology of the hunter and gatherers, the psychological qualities of openness, solidarity, oppression and boundary values and healing were prized. In posthistorical societies the rage to dominate nature creates autistic rationalizations, obsessions, paranoia, and schizophrenia--all forms of

8

_____madness. [48] However, many Third World cultures do not attempt to dominate nature and have a horrible problem with starvation. In Western Civilization recovery is difficult.

Continuum

Jean Liedloff in <u>The Continuum Concept,</u> says in hunter/gatherer culture there are no disconnections of experience between adults and children. They are constantly together. The introduction of agriculture created a loss of community, personal boundaries and archetypal issues. [49]

Addiction and Recovery

Western society has techno-addiction to machines, like the computer, television sets and missiles. Other addictions are romantic love, sex, drugs, and self-abuse. First, the way to recovery is to break denial. Second, it is necessary to come alive and break out of the mechanistic world view and its machines. Third, we should come to terms about our feelings about our lost relationships to the natural world. Fourth, connectedness is found in bounding ourselves, body and soul, with others and the earth. We should create culture that is "earth-bound, ecological and indigenous. "[50]

Gary Snyder

Gary Snyder's Deep Ecology integrates, Zen Buddhism, American Indian lore, ecology and wilderness values. Jack Kerouac used Synder as role model for Japhy Ryder, the mountain climbing poet of Dharma Bums, and a Zen practitioner. [51] Zen Buddhism is atheistic so it doesn't fit well with infinity.

Gary Synder's, Turtle Island, got a Pulitzer Prize for Poetry in 1975. Synder claims the dichotomy between wilderness and urban culture is pathological and pervasive. Thoreau claims, "In wilderness is the preservation of the world. "[52] Snyder, in response claims wilderness "is the world." To resolve the tension between civilization and wildness we must be whole. We must see ourselves "as fo od." [53]

Culture, Snyder finds relates to the Latin word, colere, meaning worship or cultivate world or liturgize it. Snyder also argues for a "culture of wilderness." Nature is the subject of science. The world is not a subject. It is within.[54] It has an objective as well as subjective dimension. Another monk master is Thomas Merton. His monastic definition because he attempts to liturgize the world is an interior Green Everevolution.

Conservation and Ritual

9

Dave Foreman is developing a new conservation movement. He is an Eco-Warrior working to change the wilderness concept from a recreational resource perception to emphasizing biological diversity. In contrast, Dolores La Chappelle maintains ritual is the means to bypass the limitations superimposed by language. For example, seasonal festivals like Mary in the month of May or the Festival of the Assumption reverse the topo for place and cosmos or the theme "the world order of a particular place" or topocosm. Non-human coinhabitants of a place, like totems follow in the footsteps of ritual. In Sienna, during Palio, the ritual horse race, all the flames of Hell are changed into paradise lights. Ritual is a connective patt ern.[55] <u>Conservation and ritual are the same movements.</u>

Eight Points of the Deep Ecological Movement

1. The well-being and flowering of human and non-human life are both crucial. They have intrinsic values independent of usefulness.
2. Richness and diversity of life forms reinforce these values and are values.
3. Humans have no right to decrease this diversity except for vital needs.
4. A substantially smaller human population is a premium.
5. Human interference or interface with the non-human world is excessive.
6. Policies affecting today's economic, technological and ideological structures must radically change.
7. Life quality should be appreciated rather than forcing a higher standard of living.
8. Those who accept the above points have a responsibility to implement changes.[56] Ecological infinity is the implication of these eight points. Cybernetics has a convergence point in infinity. Infinity is the destiny of nature of the "collective soul."

The best and most revolutionary contrast to these eight points are the Eight Beatitudes in the Bible revealed by Christ:

The Eight Beatitudes

Matthew 5: 3-11
    3 "Blessed are the poor in spirit, for theirs is the kingdom of heaven.
    4 "Blessed are those who mourn, for they shall be comforted.
    5 "Blessed are the meek, for they shall inherit the earth.
    6 "Blessed are those who hunger and thirst for righteousness, for they shall be satisfied.
    7 "Blessed are the merciful, for they shall obtain mercy.
    8 "Blessed are the pure in heart, for they shall see God.
    9 "Blessed are the peacemakers, for they shall be called sons of God.
    10 "Blessed are those who are persecuted for righteousness sake for theirs is

the kingdom of heaven.

10

By incorporating these beatitudes into the ecology movement they make the ecology movement more revolutionary than the Communist Manife st o. [57] They are the cornerstone of the right and responsibilities of the ecology movement. For example, animals are poor in spirit, need comfort, are meek, seek satisfaction, need mercy, need pure environments, need peace and suffer persecution.

Ecosophy T

Naess calls his philosophy, Ecosophy T, indicating a comprehensive "Self-realization" or "unusual symbiosis." Altruism is a consequence of this identification. He opposes the human/nature dualism which maintains civilization has transcended nature and is not subject to evolutionary/ecological laws. This is the "second nature" view. [58]

"Decreation" and Second Nature

Simone Weil, a convert to Catholicism, invented a theology called "decreation" which relates the universe to compassion and generosity, for example, the "decreative" generosity found in Christ. Martyrdom is a form of creation. Weil thought "Self-denial is a form of creation" [59] Thus, charity is a creative act which is an intuitive activity. [60] Decreation is part of any definition of the human in nature, especially ecopsychological nature. It defines the difference between first and second nature or human nature.

Modern Literature

Modern literature has an environmental consciousness expressed in the ecocentric/biological outlooks of D.H. Lawrence and Aldous Huxley. In these writers philosophical idealism, subjectivism and anthropocentrism have been superseded by "post humanism." For example, Spinoza held a nonanthropocentric outlook and a metaphysics which is found in the nature-orientated thought of Albert Einstein, Bertrand Russell and George Santayana. [61]

Modern Literature and Environmental Consciousness

D.H. Lawrence, and Aldous Huxley and Gary Snyder bring concepts like nature is unity, the world is an organism, and nature worship into modern literature. Huxley and Snyder are heirs to the transcendental tradition in Eastern mysticism. Nature has its own validity. Man is its logical partner in the whole process. This outlook is posthumanism. [62]

    D.H. Lawrence throughout his writings is on the side of the body. He thought blood and flesh were wiser than intellect. He modified this position in the <u>Rainbow</u> (1915) and <u>Women in Love</u> (1920) with a balance between the passions and intellect _____as the means to self-integration and creativity in relationships. He quarreled with the subordination of the indicators of systems of the human to the mechanical. He developed the idea of "otherness" and separation in relationships. He rejected humanism with its emphasis on man and progress. Man is part of an organic universe.[63]

Huxley favored utilitarianism which emphasizes happiness. The exchange of intuitive knowledge of the universal self would be the primary question of thought. His conception of religion is eclectic. In Island (1962) he criticizes nationalism, egoism and power lust by combining Western science with Eastern mysticism in Pala, a Mahayana Buddhist Southeast Asian country. Gary Snyder thinks technological civilization is an aberration and the depletion of fossil fuels will return man to normalcy.[64]

Zamiatin

In Eugene Zamiatin's, We, as the hero waits for his lobotomy he says, "Reason must prevail. " [65] It is an anti-utopia novel. Technology is totalitarian because "technique is efficiency" and autonomous individuals are natural and inefficient. The general or common good works to suppress individuality and the spiritual. Intellectual and emotional influences promoting dissent must further be suppressed. Integrity must be subservient to the state. Nature must have its integrity developed. Wildness threatens totalitarianism which threatens wilderness. [66] Moreover, wilderness conservationists, like the Savage in Brave New World, reject surrogates, opt for the right to claim God, society, danger, freedom, goodness, sin, unhappiness, and suffering.[67]

Forms of Ecological Consciousness

John Rodman argues there are four forms of ecological consciousness (1) resource conservation, (2) wilderness preservation, (3) moral extensionism and (4) ecological sensibilit y.[68] However, there are probably many more, for example, the collective unconsciousness, which protrudes into the consciousness. For example, the brain stem is the root of the conscious.

Human Interests

Gifford Pinchot, first, identified human interests with economic prosperity and natural power. Pinchot and Theodore Roosevelt emphasized extending the self in space and time to build a society. Rodman claims the reduction of intrinsic value to humans and the satisfaction of their interests is arbitrary, not justified, and a morally relevant quality that includes all humans and excludes all nonhumans. Secondly, maximizing value through human use leads to species imperialism. [69]

12

Ecological Sensibility

Wilderness Preservationists articulate their ideas in largely religious and aesthetic ideas. Muir, who is eclectic, claims Yosemite has value in itself. Moral extensionism is anthropocentric. Ecological sensibility has three components (1) a value theory that recognizes intrinsic value (2) a metaphysics that recognizes the significance of relationships and systems as well as individuals (3) an ethics that includes such responsibilities as noninterference with natural processes, resistance to human policies and acts that disturb the noninterference principle, limited environmental intervention to repair environmental damage, and a stage of coinhabitation that requires knowledgeable respectful and continued use of nature.[70]

The Web Metaphor

Ecology is the study of the "web of life." Paul Shepard claims the web metaphor is too simple. Ecology cannot be framed or bracketed or webbed. He claims our animal nature has long been held repellant within our Western religion and Platonism, Christian morbidity, mechanism and duality.[71] These themes form a web or internet. However, life is a more complicated mystery.

Homocentricism

Aquinas developed a homocentric, synthesis of Aristotelian and Judea-Christian dogma. In contrast, Muir thinks nature has inherent merit. Also, people are the basis of the thought of Bacon, Descartes, Hegel, Hobbes and M arx .[72] Now thinkers maintain "The rejection of animality is a rejection of nature as a whole." [73]

Hierarchies

Nature is structured after human political hierarchies. Man is omnipotent. Technology will deliver omnipot ence. [74] There are hierarchies in nature: scavengers, grazing animals and predators. Man is the highest type of predator.[75] Man is not God. Technology has not conquered man's nature, instead it has led to the widespread destruction of the earth's biomass.

Ecology: A Resistance Movement

"Ecology is a resistance mov em ent " [76] For example, ecology has its own organic essence which should brace the transience, meaning and limitation problems of man.[77] It is a movement of oppression within the hierarchies of nature and man.

Four Changes by Gary Snyder: The Transcendental Tradition in Eastern Mysticism

(1) We should halve the world population. Should we start with Gary Snyder?
(2) Control pollution from excesses and modern chemicals and poisons

(3) _____There must be balance harmony, humanity and growth with all living creatures
(4) Transformation from an urbanizing and civilizing tradition into an ecologically sensitive, harmony orientated, focus on the wild, scientific culture with a spiritual dimension.[78]

Complexity

Arne Naess claims the Shallow Ecology movement focuses on the hearts and affluence of people in developed countries and the campaigns against pollution and resource depletion. Deep Ecology has a "relational total-field image." So does the thought of St. Benedict which follows later. Deep Ecology favors a biosphere egalitarianism. It also emphasizes diversity and symbiosis. It is anti-class in theory and praxis. The fight against pollution and resource depletion requires a worldwide perspective. Complexity not complication is primary in a theory of ecosystems. For example, division rather than fragmentation is the norm. The vulnerability of life requires local autonomy and decentralization. The norms of ecology are not a product of induction or logic. Instead, life-style and ecological knowledge are suggested, inspired and fortified. [79] How does one problematize the Deep Ecology movement? The best way is through complexity. Theoretical, methodological and praxis are three levels of problematical complexity. They converge in infinity.

Ecophilosphical

Deep Ecology has normative themes and is not based completely on scientific analysis. Moreover, ecology movements are or should be ecophilosophical. Ecology is a parametered science using scientific methods. Philosophy and ecosophy require a larger perspective. Ecosophy expresses itself verbally as something with descriptive and prescriptive functions. The relation is between subsets of premises and conclusions. It is a relation of desirability. The notion of derivation must have mathematical and logical rigor, precision and use of models and systems that are tested in research.[80] Ecology and Deep Ecology should subsume and transcend science to wisdom.

Naess's Ecosophy T unites respect for the individual with respect towards ecosystems. For example, primitive cultures, like the "Circle of Life" in native American philosophy, hold one as being neither higher or lower than any other. Life is shared with the com munit y.[81] The big problem in community is the distribution of water and food.[82]

Major Ecology Themes of Great Thinkers

What are the major ecology themes of the great thinkers? Aristotle designated the hierarchical idea as the "Great Chain of Being." He thought plants were made for animals. Animals were made for man. John Stuart Mill favored a "steady state" society to oppose rapid population growth and industrialization. Darwin located the human in the web of life. Thoreau equated freedom with wildness. John Muir thought "things exist for themselves." He tried to think like a glacier. Things exist for God. During 1965 the Sierra Club, an environmental think tank, indicted overpopulation for damaging the wilderness and wildlife. David Brower believed "in the rights of creatures other than man. "[83]

Cardinal Problems and Cardinality

Rachel Carson's, Silent Spring, spoke out against pesticide use. Paul Ehrlick's, The Population Bomb, highlighted the overpopulation/environmental crisis. He helped found Zero Population Growth.[84] The way one can relate the principle of cardinality to these cardinal problems is that the world's problems should be prioritized and placed in a hierarchy. The problem sets in life are not equivalent. However, some numerical sets are equivalent. The races have equality. The principle of cardinality interfaces the Divine.

Thinker and Activist

Naess maintains it is important to be "both a thinker and an activist. "[85] Naess, in 1977, got the Sonniger Prize which is Europe's highest academic award. His works go through four periods, the first period concentrates on the philosophy of science, the second period empirical semantics, the third period is the antidogmatic period, reviving Greek pyrrhic skepticism. During the fourth period he developed Ecosopy T which stands for Tvergastein, his arctic cabin. The Deep Ecology philosophy and movement which Naess helped create has gone through three changes (1) description of the beliefs, attitudes and lifestyles of the movement (2) the creation of the Deep Ecology platform. In Ecosophy T (3) the philosophical theme is self-realization for all beings. Other themes are a classless society, complexity, diversity and symbiosis. Self-realization works for all human and nonhuman individuals. Self-realization should progress from narrow ego identification with other people to the identification of the self with nonhuman individuals, species, ecosystems and the ecosphere. He likes the ideas of interdependence and spiritual activism. Human and nonhuman life have intrinsic value but not equal value. Both have the same right to life. The genetic ontogeny of the human includes identification with all within the world. Passivity should be avoided and activism championed as Spinoza thought.[86]

Language and the Ecology Revolution

Contemporary philosophy movements which enhance the Deep Ecology position should be backed. These movements have a green linguistic dimension. For example, the Green movement compels a substantial change in language, economic, political, social and ideological structure. Also, in language, self-gift is a term preferred to self-realization. A commitment to ecology is a self-gift. The term Christian word incarnation is preferable to the word identification. Incarnation, the word made flesh, is the key to ecosophy. Martin Luther thought himself imprisoned in his body and hoped death would allow him to escape it .[87] The body is an affirmation, a word made flesh through the eucharist, rather than a prison. It is more like a sacred vessel instead of a prison. Liturgy is the way to cultivate the garden within the sacred vessel. Ecosophy and gestalt ontology states "everything hangs together" "and everything is interrelated. "[88] Thus, the ecosophy movement is also a revolution in language: green language, self-gift, self-realization, garden, incarnation, identification, imprison, all interrelate in complexity.

Treeline Metaphysics

The treeline has metaphysics. The spontaneous experience of the treeline is experience of reality which goes beyond the divisions "between subject/object and spiritual/material." Reality is fecund.[89] Joy is a key to the environment and the environmentalist and the treeline. Where is there joy in the world of fact and science? It is right at the top of the treeline. [90]

Keys to Lifestyle and Deep Ecology

(1) Simplicity
(2) Activities with intrinsic value
(3) Anticonsumerism
(4) Goods for everybody's enjoyment
(5) Opposition to "novophilia" or valuing the old
(6) Action rather than busywork
(7) Appreciation of cultural differences
(8) Solidarity with the situations in the Third World and Fourth World
(9) Lifestyles that are universal
(10)) Depths of experience rather than intensity
(11) Meaningful work
(12)) A complex positive rather than complicated lifestyle (13) Community rather than society
(14) Primary production like agriculture
(15) Efforts to meet vital needs instead of desire
(16) Living in nature rather than tourism
(17) Don't leave traces in vulnerable nature
(18) Valuing all life-forms
(19) Life-forms have intrinsic value and are not means
(20) Protection of species from dogs and cats
(21) Protection of local ecosystems
(22) The condemnation of excessive interference in nature
(23) Non-violence

(24) **Non-violent direct action**

"A Simple Path" Through the Wilderness of Life

Mother Theresa of India thinks the way of the Catholic is "a simple path" like love.[1] One can find this exemplified in the spiritual ecology of Saint Benedict who forms the best contrast to the lifestyle of Deep Ecology. Monasticism is the deepest ecology. It works against the entropy of the universe.

What did St. Benedict delineate as "a simple path" for his monastic lifestyle?

SAINT BENEDICT

"Few persons have had a more significant influence on the spiritual lives of others than St. Benedict of Nursia, Patron of Europe and found of an illustrious Order of religious.

What advice does he give those interested in living out the Gospels? How does he interpret and apply the message of Jesus? Do you wish to know and follow the guidelines to sanctify that have proved immensely successful? Here they are --

1.   In the first place, to love the Lord God with the whole heart, the whole soul, the whole strength.
2.   Then, one's neighbor as oneself.
3.   Then, not to murder.
4.   Not to commit adultery.
5.   Not to steal.
6.   Not to covet.
7.   Not to bear false witness.
8.   To respect all men.
9.   And not to do to another what one would not have done to oneself.
10.   To deny oneself in order to follow Christ.
11.   To chastise the body.
12.   Not to become attached to pleasures.
13.   To love fasting.
14.   To relieve the poor.
15.   To clothe the naked.
16.   To visit the sick.
17.   To bury the dead.
18.   To help in trouble.
19.   To console the sorrowing.
20.   To become a stranger to the world's ways.

21. To prefer nothing to the love of Christ.

22. Not to give way to anger.
23. Not to nurse a grudge.
24. Not to entertain deceit in one's heart.
25. Not to give a false peace.
26. Not to forsake charity.
27. Not to swear, for fear of perjuring oneself.
28. To utter truth from heart and mouth.
29. Not to return evil for evil.
30. To do no wrong to anyone, and to bear patiently wrongs done to oneself.
31. To love one's enemies.
32. Not to curse those who curse us, but rather to bless them.
33. To bear persecution for justice sake.
34. Not to be proud.
35. Not addicted to wine.
36. Not a great eater.
37. Not drowsy.
38. Not lazy.
39. Not a grumbler.
40. Not a detractor.
41. To put one's hope in God.
42. To attribute to God, and not to self, whatever good one sees in oneself.
43. But to recognize always that the evil is one's own doing, and to impute it to oneself.
44. To fear the Day of Judgment.
45. To be in dread of hell.
46. To desire eternal life with all the passion of the spirit.
47. To keep death daily before one's eyes.
48. To keep constant guard over the actions of one's life.
49. To know for certain that God sees everyone everywhere.
50. When evil thoughts come into one's heart, to dash them against Christ immediately.
51. And to manifest them to one's spiritual father.
52. To guard one's tongue against evil and depraved speech.
53. Not love much talking.
54. To listen willingly to holy reading.
55. To devote oneself frequently to prayer.
56. Not to fulfill the desires of the flesh; to hate one's own will.
57. To obey in all things the commands of the church.
58. Not to wish to be called holy before one is holy; but first to be holy, that one may be truly so called.
59. To fulfill God's commandments daily in one's deeds.
60. To love chastity.
61. To hate no one.
62. Not to be jealous, not to harbor envy.

63.   Not to love contention.
64.   To beware of haughtiness.
65.   And to respect those who are older.
66.   To love those who are younger.
67.   To make peace with one's adversary before the sun sets.
68.   And never to despair of God's mercy.

These, then are the tools of the spiritual craft. If we employ them unceasingly day and night, and return them on the Day of Judgment, our compensation from the Lord will be that wage He has promised: Eye has not seen, not ear heard, what God has prepared for those who love him. "[2]

The ecology movement is a spiritual craft and these are its tools. They should be employed unceasingly day and night throughout all of the world's ecosystems. Carl Jung thought the world is a monastery. St. Benedict's monastic rule is an internal Green Revolution or spiritual revolution within the monastery of the world. His lifestyle is "a simple path," or focus within the entropy in the universe.

The New Age Movement and the Green Party Movement: Father Chardin and Petra Kelly

Pierre Teilhard de Chardin inspired the New Age Movement along with Buckminister Fuller. They are working for the humanization and humanity's takeover of the Earth. [1] In contrast, Petra Kelly, an Irish-German American, masterminded the Green Party Movement in Germany. She died recently in an insane act of murder suicide. Was her death a comment on the direction of the ecology movement? This may be true. However, Patricia Ireland thinks she just needed a rest. [2]

To study Chardin's tradition, John Passmore traced the Western views of human superiority and dominion of the Earth to Platonic/Christian stewardship and the idea of human perfecting nature and the change from "first nature" to second nature" which is separated and alienated from the Earth. First and second nature are fallacious typologies. Aristotle began this evolution Bacon/Hegel/Marx and Chardin continued it. New Age thinkers and Social ecologists think humans should direct evolution. Thomas Berry rejects Chardin's anthropocentrism and misguided technological utopianism. Are "space colonies," like Disney World on earth, similar to a modern Plato's Cave. [3] The answer is largely yes. For example, the basement of the Mall of America in Bloomington, Minnesota, is a modern Plato's cave. Plato's cave may also be a sexual symbol.

In contrast to Chardin, Deep Ecology favors ecocentric egalitarianism. A central ecofeminist premise is to counter androcentrism which is a male centered universe. In contrast to androcentrism, the "anarcho-socialist" position advocates decentralization and opposes hierarchy. Also, Oriel Sallen thinks women should love what we are and "flow within the system of nature. "[4]

The Intellectual Godfather of the New Age: Father Chardin

Chardin has an "anthropocentric progress-orientated evolutionary cosmology." He thinks all matter has consciousness which provides for greater complexity. Evolution moves "from non-living matter to single-celled life, and then to complex organisms." Evolution is moving towards more reflective brain power. Side shows of evolution like plants or insects are insignificant. Human enterprise and thought converge in an Omega or Center of Attraction. Human technological advance is inevitable. "The Earth is Man," thinks Chardin. Man is evolution. The earth will be covered by the "arch-molecule" of humanity with its consciousness. We live in an artificial environment, megatechnology, an artificially created "second nature. "[5] The writer thinks the ecology movement needs a different Godfather than Chardin, like God, the Father.

Chardin's "New Age Disembodied Cybernetics"

Chardin maintains the noosphere is a convergence movement towards the human species which reverses the evolutionary flow towards diversity and divergence. The noosphere is based on technology and communications and the transmission of civilization to the planets. Nature is subsumed into human control. Technology and genetic manipulation will manage nature. Greater population will further actualize the noosphere. Cultural diversity will lead to global conflict within the noosphere. The Omega Point and global monoculture will produce planetary culture. By pursuing Chardin's noosphere to the extreme we have "new age disembodied cybernetics," such as computer consciousness."[6]   However, there is a metaphysics to all organisms. For example, God cares for the sparrows in the Bible.

St. Benedict and St. Francis

St. Benedict thought we should recreate paradise out of the chaos in the wilderness. For example, the Christians are stewards on the farm of the Earth. This farm is to be altered from its chaotic wild state. St. Benedict argues for man over nature or man over entropy in modern terms. St. Francis calls for "equality of creatures." Chardin is part of the man perfecting nature tradition.[7] Many cultures do not have the concept of stewardship and dominion over the farm of the earth. Massive starvation, illiteracy and medical horrors stories are often the result. The rejection of the idea of stewardship is a form of intellectual leprosy.

Wildness and Wilderness

The Deep Ecology Platform is committed to the protection of wildness and wilderness regions. Wilderness protection should go beyond anthropocentric considerations such as aesthetic "frontier" virtues, recreation, character development and "spiritual" values to include protection of forests and biodiversit y.[8] Aquinas O'Dougherty thinks "people are problems. "[9] In contrast, Reverend Neal Thompson thinks people are "challenges. "[10] Wilderness is both a problem and a challenge for people.

Wilderness areas have ecological relations. Thoreau claimed "all good things are wild and free." Centuries of life as primitive beings developed a genetically founded "Paleolithic human nature." Wilderness is not a commodity or a business like wilderness tourism. The wilderness areas have been imprisoned. Global ecosystem zoning to protect ecosystems and biodiversity have been proposed. For example, Naess thinks the earth should contain "one-third wilderness, one-third free nature, and one-third bioculture." As civilization postevolved and evolved "out of nature" to create a "second nature" or dualism. The wise use movement attempts to eliminate wilderness areas for the earth's total exploitation.[11]

Ramachandra Guha: Social Ecologist

A Social Ecologist from India, Ramachandra Guha sees the two largest ecological problems on the globe as overconsumption by the first world and Third World urban elites, overpopulation, ozone layer depletion, habitat loss and loss of species through extinction. He favors emphasis on the human species for social justice and equity. Society should be restructured to achieve sustainability, social justice, steady-state economics and radically shifted rates of consumption and production habits. The first world should embrace antipoverty movements, alternative lifestyles and relevant technology .[12] However, "Man's problems are man made. "[13] Cows eat more than the people do in India.[14] This is a major reason for the starvation in India.

The Stewardship Approach Versus the Complexity of Nature

The stewardship approach maintains restoration projects should be like agriculture farms enhanced by people. Scientific predictors are an important factor in this approach. Some ecologists, in contrast, claim nature is more complex than their own thinking capacity. In contrast, some think nature's preserves should be static and free from manipulation? Should it be allowed to take its course? For example, free nature is often sparsely populated. Many thinkers maintain this state should continue. Also, a bioregionalization of garden scenario may be in the works because of the population problem.[15] The stewardship approach versus the complexity of nature debate continues.

Biomass and Bioculture

The six million humans have destroyed 40 percent of biomass productivity on earth. A destruction between vital and nonvital needs and wants is necessary. Bioculture is the culture humans create to regulate and exploit the environment such as pet farm s.[16] However, both biomass and constructive bioculture are vital for the earth.

Are there contradictions between biomass and bioculture, nature and culture, between wilderness and sustainable development? Many ecology thinkers maintain this.[17] The concept of limits differentiates or finely distinguishes these concepts. Humankind should know and define their limits.

Neocolonialism

Is Deep Ecology in the Third World a form of "neocolonialism."[18] The answer is yes and no. For example, instead of sustainability, Naess uses the phrase "wide ecological sustainability" within the context of complexity. Thus everything, not just politics, potatoes and society needs to be revolutionized. Life sustaining agriculture is not a form of neocolonialism. Western "cultural garbage," for example much of pop culture, is.[19] To deal with local problems Gary Snyder proposes a world tribal council unifying cultural and individual pluralisms. [20]

Truth: A Blender?

Particularities and pluralisms characterize life. Truth is not usually the product of a blender. The blender idea may be a form of neocolonialism. A world symphony with orchestration is a better metaphor, for example, the world symphony of ecology. Symphonies have a metaphysical dimension like food--nectar of the Gods. Food also expresses beauty which is universal. Maybe the West needs to be colonized by the symphony of the Third World.

Diversity, hope, activism, and learning are the keys to the Green societies of the future and to truth. The writer maintains learning is a commandment of the Green movement. Also, the Pope is its intellectual vicar, papa and steward.

Father Hilary Freeman .

Father Hilary Freeman addresses man's harmful effects on the globe. He looks at systems analysis and computer simulation in a letter dated April 8, 1996. In a later undated letter he champions control of nuclear w ast e. [21] Father Hilary Freeman is the St. Thomas Aquinas of the ecology movement. Both Freeman and Aquinas are Dominicans.

Taxes

Opposition to taxing nonprofit corporations, like the Catholic Church and the International Alliance for Sustainable Agriculture, are also important. Greenpeace is also a nonprofit. Aquinas O'Dougherty thought "Government is the original original sin. "[22] John Dewey thought "big business is the shadow government casts on society." Government is often the problem not the solution.

Oriental Ecology: Hindu Bardos and Christian Stewardship

What are the consequential relationships between Hindu reincarnation bardos? Mike Franey thinks if one can never explain the consequential relationships between the transition state and reincarnation, then it just is not true. Christian stewardship, then, is the major scientific alternative. For example, Catholicism looks to Aristotle the father of Western Science to explain consequential relationships.

Axioms of Ecology?

"Most animals die violently or starve to death. "The environment reinforces aggression" for fo od.[23] "Utopia is the Greek word for no place. "[24] Utopia is not a

Never before has man had such capacity to control his own environment, to end thirst and hunger, to conquer poverty and disease, to banish illiteracy and massive human misery. We have the power to make this the best generation of mankind in the history of the world--or to make it the last.

President John P. Kennedy, address before the General Assembly of the United Nations, New York City, September 20, 1963.--Public Papers of the Presidents of the United States: John F. Kennedy, 1963, p. 696.

C Copyright 1996 by the Hellenist International Institute Publishing Co.

The Hellenist International Institute Publishing Company
A Royalist\Radical Publishing Company
Riverside Plaza M3410
1615 South Fourth Street
Minneapolis, Minnesota 55454

O'Dougherty, Patrick Aquinas, Ph.D., mock Nobel Laureate. 1946--

St. Patrick, The Green Revolution, The Hydrogen Conversion Project

A Spiritual Dedication to Francis Duff and the Legion of Mary and to the Oblates of St. Benedict

An Intellectual Dedication
To Mike Franey, David Noble, Mark McGee, Brother Gregory Conant, Father Hilary Freeman, Sister Mary Anthony Wagner, and the Oblates of St. Benedict at St. Benedict College, St. Joseph, Minnesota and St. John's University, Collegeville, Minnesota.

Family and Friends
To Richard and Patricia O'Dougherty-Kast. To my father Aquinas O'Dougherty, my brother Mike and my sisters: Margaret Mary, Mary Ann and Maureen. To Megan, Sean and Stephen. To John O'Dougherty, John O'Dougherty, John Herlick, the Conant family and the Linstroth family

ISBN: 0-9626665-7-2

Library of Congress Catalog Card Number: 96-94777

Printed in the United States of America

TABLE OF CONTENTS

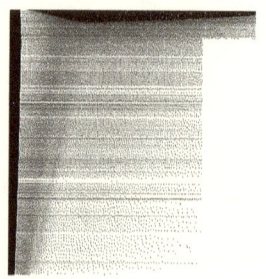

BIBLIOGRAPHY

ABBREVIATED INDEX

97

101

possibility. Can the field of ecology be axiomatized? These may be just a few of its axioms. The question is are they self-evident?

David Noble and a Return to Orthodoxy

David Noble favors a return to Episcopalian Orthodoxy. For example, he might argue the ecology relationships are essentially covenantal relationships. Stewardship and dominion are covenantal relationships. Is ecology literature biblically apocalyptic. He produced a Ph.D. by Christopher Lewis entitled, "Progress and Apocalypse: Science and the End of the Modern World," which studied this type of literature which weighs highly on the negative.[25] The literature of ecology is a literature of hope as well. Faith, hope and charity are "the simple paths" and keys to ecology .[26] Arm oneself with hope! Mysteries, like faith, hope and charity rather than apocalyptic visions are the best visions for the species and the races.

25

APOCALYPTIC VISIONS

What are some apocalyptic visions of threats to the species and the deconstruction of the races? In Albert Camus' novel, "The Plague," he writes that pestilences recur in the world. Yet we find it difficult to believe in the ones that crash down upon us from the sky. Laurie Garrett in her book, "The Coming Plague," depicts a frightening picture of many infectious diseases attacking mankind such as AIDS, cholera, tuberculosis, yellow fever which are infectious diseases. There are many newly emergent viruses and bacterial infections such as Legionnaires' disease, toxic shock syndrome, Lassa fever and others as Ebola and Marburg viruses.[1]

Ms. Garrett's assessment is gloomy because of humanity's underestimation of microbes which she claims are winning. She claims many developments have amplified the range and strength of microbes resistant to most treatment, for example, penicillin-resistant staphylococcus, acyclovir-resistant herpes and AZT-resistant H.I.V. Pesticides lead to declining diversity in the insect world. The building of large dams in Egypt, Ghana and the Sudan has led to increasing numbers of the schistosome parasite. Deforestation shifts the microbe population. Urbanization in the Third World has created prime conditions for epidemics. Government is often the problem rather than the solution. The AIDS crisis is mainly the top of the microbial iceberg. [2]

Extinction

Charles Birch claims the world is a Titanic. For example, the relationship between chlorofluorocarbons and ozone depletion could prophecy the world's end. Scientists have conjured up an "impossibility theorem" which maintains the high rate of consumption and pollution by wealthy nations like America and Japan is not possible for all the peoples of the earth. Is inequality in wealth justifiable? The answer is yes. The Church teaches some class differentiation promotes talent and excellence. Birch argues worldwide industrialization would be lethal to the earth. He blames the rich. 1,000 species annihilation a year equals a "holocaust of nature." A biocentric ethic that perceives in all life an intrinsic <u>and</u> a instrumental value is a categorical imperative. [3]

The Deconstruction of the Races

If the species start to disappear, the tribes are next, then the human race may delete itself. The races are being deconstructed through the problems facing ecology and society. And "Man is not, as he may think, ruler of the universe." Quoting Father Hilary Freeman. Man is not the ruler of the races either.

A Catholic, Scientific Conceptual Framework and Methodology

There has been a massive intellectual contribution to mathematics and science made in the twentieth century. Catholics should create a new Catholic scientific revolution. For example, one can take a subdivision of a field, like sustainable agriculture, in ecology and apply it scientifically using a methodological technique to analyze agriculture in biblical texts or in Catholic theology. This can be done for thousands of new areas in research in all levels of science and mathematics. The idea is to invent new fields like the personal intuitionist school of ecology which follows, then apply them to theology in a manner consistent with Catholic teaching with hope rather than apocalyptic visions.

27

CATHOLIC PERSONALISM: AN INTERPRETATIVE APPROACH TO THE GREEN REVOLUTION

Ecology themes relate to the writer's Catholic personalist/intuitionist school of mathematics which is a powerful interpretative approach to ecosophy. Personalism is "any doctrine or movement which emphasizes the rights and centrality of the individual being in his social, political, intellectual, etc. milieu."[1] Richard Kast relates personalism to love, affection and trust versus American narcissistic encounters and American matter of fact relationships. The word, "good," relates etymologically to Anglo-Saxon word for God who is part of any definition of personalism, interdependence, counterpoint, and mathematics.[2] Ecology is "good" personalism.

Brother Gregory Conant, O.S.B., On God and Personalism

Conant, O.S.B., relates personalism to God. He contends God is "self-gift" and love. Conant, O.S.B., says this self-gift creates the complexity in God and in everything else. He thinks a one person deity cannot have social value and be eternally. This indicates God must not be just one person but at least two. Conant, O.S.B., argues, "to be love, the two must enjoy a generosity beyond mere self-interest, which shows there must be at least three. "[3] The Holy Spirit is "the Principle of the free service of others," which is the "nature of God. "[4] Christian ecology has an eternal social value and virtue eternally.

The Genesis of the Universe

The personal intuitive unites the personal with **activities,** like creation. Conant, O.S.B., suggests the personal intuitive begins with Genesis 26-27: "Let us make man in our image, after our likeness; and let them have dominion over the fish of the sea, and over the birds of the air, and over the cattle, and over all the earth, and over eay creeping thing that creeps upon the earth." So God created man in his own image, in the image of God he created him; male and female he created them. And God blessed them, and God said to them, "Be fruitful and multiply, and fill the earth and subdue it; and have dominion over the fish of the sea and over the birds of the air and over every living thing that moves upon the earth."[5] The Bible is a proof of intuitive personalism. It also contains myriad ecological proofs which relate to intuitive personalism such as the dominion theme which is part of the definition of stewardship. Genesis is deductive personal activity truths.

A Definition of Mathematics

Mathematics is a language, especially of science. A language is "a special set of symbols, letters, numerals, rules, etc. used for the transformation of information, as in a computer." [6] A science is "a branch of knowledge or study esp. one

28

concerned with establishing and systematizing facts, principles, and methods as by experiments and hypotheses." In particular mathematics deals with "the group of sciences (including arithmetic, geometry, algebra, calculus, etc.) dealing with quantities, magnitudes, and forms, and their relationships, attributes, etc., by the use of numbers and sy mbols" [7] However, mathematicians cannot agree on a definition of the field in part because it has a spiritual, personal, undefined and <u>intuitive</u> dimension.[8] **Intuitionists, like Jan Brouwer, think mathematics is activity.** For example, it is a <u>pregnant</u> methodology. A methodology is "the science of method, or orderly arrangement; specif., the branch of logic concerned with the application of the principles of reasoning to scientific and philosophical inquiry."[9] **A methodology is also an activity, like pregnancy.** Also, evolution is a necessary element in any definition of mathematics. Is personalism mathematics? Yes, mathematics applies as the language and methodology of a philosophy which defines the person, for example, an evolving person in a ecosystem.

Female Intuition and the Intuitionist School of Mathematics

Women have a higher sense of intuition than men--a sixth sense or female intu1t1on. Intuition is "the direct knowing or learning of something without the conscious use of reasoning," and "the ability to perceive things without conscious reasoning."[1] Intuitionism "is the doctrine that all things are apprehended in their real nature through intuition. "[2] In ethics it is "the doctrine that the rightness of acts or fundamental moral principles are apprehended by intuition."[3] Brother Gregory Conant contrasts ratiocination or formal logic with the "intuitive: an immediate higher form of knowledge. "[4] Also, intuition has religious dimension. For example, there are personal, intuitive realities. Plato thought, for example, that all knowledge did not come through the senses. There is intuitive truth within the mind. Ecological truths, for example ecopsyhology, are part of these intuitive truths within the mind and brain stem.

An Intellectual Biography of Brouwer: The Founder of the Intuitionist School of Thought in Mathematics

How do mathematicians define the intuitionist school? The key is the question of existence. Jan Brouwer disagreed with Poincare on mathematical existence. Brouwer did not think mathematical existence meant <u>freedom from contradiction</u> as Poincare thought but **intuitive constructibility.** Brouwer conceived of mathematics as a **free ("mind")** building mathematical objects beginning from **self-evident primitive intuition.** The masculine and feminine free "minds" should redefine primitive intuition. Do masculine and feminine thoughts have a self-evident basis? In contrast, formal logic is a means of describing regularities in systems already built. Formal logic lacks value in proving the foundations of mathematics. The absolute validity of logical principles is questionable. The axiomatic foundation of mathematics failed. Brouwer intended that David Hilbert would be unable to prove the consistency of arithmetic.

Even if Hilbert succeeded, Brouwer thought this would insure a mathematical system defined by axiom s. [5] **The mind of men and women is free ("mind") building from primitive objects of intuition. "** [6] **Does it lack consistency? Is it free from contradiction? The answers are found in the primitive. This also applies to the problems in the intuitive algebra of ecology, for example, relationships throughout nature. The keys to the ecological mind are** primitive **objects of intuition. For example, wilderness is the primitive.**

Catholic Personalism
Karol Wojtyla (Pope John Paul II) Personalism and Intuitionism
Wojtyla's Conception of the Person Defined

"A person is an objective entity, which as a definite subject has the closest contacts with the whole (external) world and is most intimately involved with it precisely because of its inwardness, its interior lif,a." [7] Intuitionism or activity is part of the definition of the person because activity is part of the objective as well as interior life of the person. Could the person also be an incomplete definition for the field of mathematics? The answer is yes. Mathematics has objective and intuitive personal dimensions. **Catholic personal intuitionism is a basis of ecological nature.**

Ecology is primarily theology. Ecology is personal. The truth of ecology is spiritual. This writer founded a new school of ecology, ecological spiritual intuitionism. Mystery is the basis of the universe and of this new school of ecology.

30

PROJECT HELIOS: THE HYDROGEN CONVERSION PROJECT--A FOCUS

Helios: "the ancient Greek god of the sun, the son of Hyperion and Thia, and father of Phaethon: represented as driving a chariot across the heavens. " [1] The sun is largely fueled by hydrogen reactions.

A preface to the hydrogen conversion project by a letter written by Brother Gregory Conant OSB.

Dear Patrick,   March 6th. 1996
The thing I know is that the Hydrogen will be taken out of H20 and you may be pleased to learn that my initial worry about the water supply on Earth has been allayed by a movie about the weather which reported ecologists are worried on account of the fact that melting of the polar ice caps is deepening the oceans so that if it continues it will inundate half the World's great cities since they are less than a hundred feet above sea level.[2]

An additional advantage is naturally the release of oxygen into the atmosphere even if a little of it might be bottled for medical or industrial use. Indeed, it might become inexpensive enough so that people could buy it to improve the atmosphere in their homes or other limited spaces to draw greater good out of it than could come from its direct release into the atmosphere. There is no telling how greatly it might improve the quality of life indoors, especially during the winter when it might reduce the incidence of respiratory ailments. You should perhaps fill a chapter with these advantages and entitle it FRINGE BENEFITS.[3]

I believe to get hydrogen fuel going you need a government in a high tech country having no immediate access to oil in its territory. This says to me at once that you ought to have your book translated into Japanese and send copies of it to both the related government offices and the different departments of the universities such as economics and engineering which would have the people most likely to want to produce the results you wish. Once some country has achieved the change to avoid being behind. Several other countries which might want to initiate the change are

Korea, China, England, France, Germany, Poland, and Israel: in order to cut the economic oriental rug from under the Arabs.[4]

It seems to me that industrial plants for the reduction of seawater for hydrogen fuel could produce most other products of desalinization as well. And that the plants producing hydrogen fuel could produce their own electric power and that additional electricity produced and sold could make them quite profitable. Still, there may be places where mobile generators are required like a starter motor on a diesel engine or flint and steel used for lighting a gas flame, to start things off at the beginning. But I should also tell you that petroleum is worth much more in plastics than in fuel so that to reduce the amount used in fuel is good because it increases the supply available for plastics.[5]

Brother Gregory Conant, OSB

Definitions and Classical Themes

Hydrogen: In Chemistry it is a "colorless, odorless, flammable gas that combines chemically with oxygen to form water: the lightest of the known elements. "[6]

Sea: "the salt waters that cover the greater part of the earth's surface. "[7]

Ocean: "the vast body of salt water which covers almost three fourths of the earth's surface. "[8]

Neptune: "the ancient Roman god of the sea, identified with the Greek god Poseidon. "[9]

Mariology: Mariology relates to hydrogen because Mary and hydrogen are the pure elements or known for their power and purity. There is strength in purity. There is strength in grace. Also, the writer intends a pun between sun and son.

Hydrogen in Europe

Berit Pegg-Karlsson is the Swedish-born leader of the British-Scandinavian Association for Wind and Hydrogen Power. John Seymour, the self-sufficiency writer, is the main trustee of the Pure Energy Trust which endows this association. Pegg-Karlsson is popularizing the hydrogen 'Welgas' experiment which the Swedish steel industry, SAAB, and other firms finance in the town of Harnosand. Olaf Tegstrom, in Harnosand, designed a house where the electricity starts from a small Danish windmill in his garden. The electricity is computer controlled. The electricity electrolyzes filtered water into hydrogen and oxygen which are its constituents. The hydrogen gas heated the house and fueled the stove and his SAAB car. This car does not pollute because the exhaust consists largely of water vapor. Safe storage is not a problem because the gas absorbs to make metal hydride and releases as needed. In West Berlin, Daimler Benz has created a filling station where converted vehicles refuel with hydrogen created from town gas. [10]

The Hydrogen Conversion Revolution

Hydrogen will become the world's prime energy. This technological revolution will replace nuclear power and solve the atmospheric pollution problem. Hydrogen has an energy content three to four times higher than oil. It originates from all known energy sources. It is a by-product of many industrial processes. [11]

American Hydrogen Powered City

32

An American city linked to other hydrogen cities to share ideas and head joint hydrogen projects is an imperative. Minibuses powered by hydrogen produced by a wind turbine and an electrolysis unit are on the technological horizon. Hydrogen power is Jules Vernes' ancient dream come true, that is, water is a powerful hydrogen fuel source for farms and cities. [12]

1p per mile: A Hydrogen Car

Dr. Roger Billings, who invented the first home computer and the double-sided floppy disk, patented a Laser-Cell-TM fuel cell. This hydrogen fuel cell extracts hydrogen from water from electric mains to which it is plugged for approximately eight hours. It stores the gas in powdered metals so it cannot explode or ignite. [13]

The fuel cell can switch into 'reverse' and turn the hydrogen into electricity which powers the electric motor driving the vehicle, for example, buses, trains, and space craft. The cell has no moving parts and is one-third the size of a Fiesta gas engine. This cell requires no service and a life span of over 250,000 miles. The fuel-to-power ratio of the gas engine is thirty per cent. In contrast, the fuel-to-power ratio of hydrogen converted to energy is 60 to 80 per cent. The hydrogen prototype car is 'fast and quiet' and costs less than 1p per mile to operate. It is environmentally friendly. The exhaust is water vapor. Two gallons of water create enough hydrogen for a 300 mile trip. The car speed tops 80 mph. Dr. Roger Billings is the Director of the Academy of Science, Kansas, USA. [14]

The Hindenburg Blimp Crash

The Hindenburg was a German dirigible. It was the greatest rigid airship built and a victim of a spectacular crash. The Hindenburg was a 804 foot long airship. This design was first launched in April, 1936 at Friedrichshafer, Germany. Its maximum speed was 84 miles per hour. On May 6, 1937, the hydrogen inflated Hindenburg, while landing at Lakehurst, N.J., burst into flames and crashed. Thirty-six of its 97 passengers were killed. The fire was blamed on a discharge of atmospheric electricity near a hydrogen leak. It may have been the victim of anti-Nazi sabotage. Also, the Nazis may have intentionally crashed the Hindenburg to dramatize the failure of Americans to sell the Germans helium to produce the necessary hydrogen fuel. [15] Helium could refuel the Nazis' war machine. [16] Hydrogen is a perpetual fuel.

The Fuel Water Cycle

Hydrogen comes from water in a perpetual fuel water cycle. Water is the main byproduct when hydrogen burns. In 1776 Henry Cavendish, the British chemist, found hydrogen by dissolving metals in dilute acids. Antoine Lavoisier, the French Chemist, in 1783 named hydrogen from the Greek words saying "water producer." Electrolysis is the creation of hydrogen by passing electric current through water.

33

Besides water, hydrocarbons, a class of chemical compounds, contain hydrogen and carbon. Also, burning hydrogen in a fuel cell produces electric current .[17]

In Peter Hoffman's, "Forever Fuel," the advantages of hydrogen summarize as follows:

(1) Hydrogen is a perpetual fuel.
(2) It has a variety of combustion methods.
(3) Hydrogen combustion creates nitrous oxides, low level pollutants, which can be eliminated by combustion control methods.
(4) Hydrogen has about three times the per unit weight of gasoline. It is the most energy per unit weight of energy fuels.
(5) Hydrogen can be moved through pipelines.
(6) It is nontoxic.
(7) It dissipates rapidly in air minimizing explosion problems.

The disadvantages of hydrogen are as follows:
(1) Hydrogen burns at lower concentrations than other fuels. Thus it has a higher range of flammability.
(2) Hydrogen storage is more complex than liquid or gaseous alternative fuels.
(3) Hydrogen's low liquefaction temperature (-453 degrees F) demands a large energy input for refrigeration.
(4) On a volume basis hydrogen has a low energy content. Thus compared to gasoline its storage is bulky.
(5) Hydrogen has special safety problems because of its high flame velocity and low ignition energy.[18]

The Intuitionist Framework and the Hydrogen Atom

The activity of the hydrogen atom is the dominant activity in the universe. Motion does not stop until absolute zero occurs. Constructible equations for the hydrogen atom and its exponential number of interactions are intuitively constructible.[19]

Statistics

The United States produces nearly one billion cubic meters of hydrogen every year in the United States. The figure does include quantities used in refining electric current. The combustion of hydrogen gas can produce electricity as well as heat energy. Hydrogen composes 90 percent of all atoms in the universe. In interstellar space there is an average of one hydrogen atom for each cubic centimeter. [20]

Production

34

Ammonia, methanal and synthetic gas all contain hydrogen. The most common method of creating hydrogen is to take it from hydrocarbons like methane, fuel oil, gasoline and crude oil. In contrast, electrolysis creates hydrogen, in excess of 99.9% pure, but it is expensive. It is a process of producing hydrogen and oxygen from an electric current passing through water which breaks down the chemical bond, the product is two positively charged hydrogen atoms and a negatively charged oxygen ion.[21]

Electrolysis: Synthesis

Electrolysers composed of four elements: a container, an electrolyte, an anode, an cathode and a separator. The electrolyte contains water and a chemical inserted to allow the conducting of current. The chemical can be a salt, a base or an acid. The current moves between the electrodes. The separator is set between the electrodes. It allows the current to pass. [22] However, a basic problem in creating electrolysis is to let ions travel between the electrode while at the same instant excluding gas molecules. Solid porous separators used with liquid electrolytes keep gasses separate but allow ion transfer. Moreover, there are advantages and disadvantages to parallel compared to series electrolysers. High temperature electrolysers provide increased performance. [23]

Hydrogen is seen as a source of energy like petroleum or coal. However, hydrogen does not occur naturally in large quantities. Most of it is comes from steam heating hydrocarbons and water or through electrolysis. Energy from sunlight changes to electricity by photovoltaic cells (PV). <u>This electrical energy is held in batteries or as hydrogen through electrolysis.</u>[24]

Production of Hydrogen by Chemicals

Chemicals can produce hydrogen. For example, hydrochloric acid is the most suitable electrolyte when added to water in an electrolyser. Also, sulfur-assisted water electrolysis is another probable process for the co-generation of hydrogen and sulfuric acid.[25]

Other Methods of Hydrogen Production

Photovoltaic Processes and electrolysers can combine in a single device. Also, some hydrogen compounds are simpler to split than water, for example, hydrogen sulfide. Biological sources such as carbon-based plants or municipal sewage or algae can produce carbon. Specific bacteria, for example, will change organic matter into methane. At 2,730 degrees C water changes into hydrogen and oxygen. A parabolic lens focusing solar energy can reach this temperature. [26]

Turbines and Thermochemical Cycles

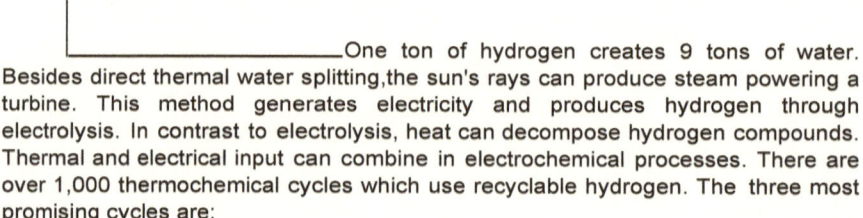

One ton of hydrogen creates 9 tons of water. Besides direct thermal water splitting,the sun's rays can produce steam powering a turbine. This method generates electricity and produces hydrogen through electrolysis. In contrast to electrolysis, heat can decompose hydrogen compounds. Thermal and electrical input can combine in electrochemical processes. There are over 1,000 thermochemical cycles which use recyclable hydrogen. The three most promising cycles are:

(1) The Bismuth sulfite cycle, starting at Los Alamos Scientific Laboratory.
(2) The sulfur cycle at Westinghouse.
(3) The iodine-sulfur cycle created through the General Atomic Company in joint research with the Joint Research Center at Ispra, Italy.[27]

Recycling Metric Oxide Acids and Compounds

Newek C. Cook created a process for recycling metric oxide acids and compounds while splitting water to create hydrogen. Nitric oxide has a low boiling point, higher thermal stability and a low ionization potential. However, the hydrogen yield is low. These problems are extremely complex from a compound technical viewpoint. [28]

Fuel Converted From Trash

Steam reforms coal. Hydrogen comes from steam reforming many carbon-containing materials such as trash. An "exothermic reaction" or a reaction using a source of heat such as reacting carbon with steam creates hydrogen and carbon monoxide. The resulting mixture of hydrogen and carbon monoxide is "town gas." It was in street lighting in 1810 in London. It spread to Boston and New York by 1830. It was a cleaner fuel than coal or oil for lamps.[29]

Steam Reforms Biomass

Part of the cost of creating hydrogen can come from converting organic waste or biomass to hydrogen fuel. Income results from disposing waste material and selling a high grade fuel. The carbon monoxide arising from a steam reforming operation can burn in air to recover part of the energy. The heat of combustion can supply energy for a steam-reforming reaction. The oxidation of hydrocarbons makes about three times by weight the carbon dioxide as found in the original hydrocarbon.[30] The interpretative framework of intuitionism analyzes the personal process of converting organic waste to hydrogen fuel. This systemic process has intuitive constructibility. The activity is programmable and mathematically constructible.[31]

Pyrolysis

Pyrolysis is the heating of hydrocarbon to several hundred degrees without oxygen to extract pure carbon. This pure carbon changes into many carbon-containing materials such as activated carbon filers, high strength fibers for car bodies, fishing rods, rust proof plastic buildings substances, artificial diamonds, electronic circuit platings, Buckminister Fuller, "Buckeyballs," and hydrogen storage. [32] These carbon-containing materials are quite versatile. Pyrolysis has intuitive constructibility.

Hydrogen Paradoxes

More garbage forms than coal in the United States each year. They have a similar carbon content. There is more carbon wasted than mined.[33]

The Storage Problems of Hydrogen

There are three main methods of storing hydrogen: gas, liquid, or solid hydrogen. Hydrogen stores as a gas in containers with high pressure compound of storing, light-weight materials. Fibers such as aramide, glass or carbon are wound around the container to contain leaks. Hydrogen emits from the container through high pressure. Regulators are necessary to decrease the pressure before the fuel injects into the engine. [34] The storage problems of hydrogen have mathematical intuitive constructibility.

Hydrogen liquifies at --423 degrees **F.** The liquid hydrogen storage container must have insulation. A special pump supplies the fuel to the engine. [35]

Hydrogen in gas form stores in solid (hydride) storage, for example, metal. The metal absorbs the gas at both high and low temperatures. Heat and low pressure releases. Water surrounds the hydrogen from storage tubes in a liquid hydride configuration. Heating or cooling the water removes the hydrides in the tubes.[36]

Definition of a Hydride

What is a hydride? Hydrogen stores in a "solid"fotrhi ry.etals like magnesium, titanium and iron which absorb hydrogen when cooled and release it through heat. It remains a gas but imperceptibly confines in the me}1} !'1.1£1'¥ t9 the molecules. If the metal charges with hydrogen gas, it is a h ydride. [37] s . .Cf•. C:

Hydrides and Stora "[{§;(J(rf ,\i/

hydrogen has more density than H,y1t   ' IpeEJdS expensive
storage containers and intricate refueling procEl?!i .····   n eded to liquify
hydrogen gas equals about one-third of the ener(J.\I.' •
in hydrides the total volume and weight of the.hx   '<
                                                              'In
                                                          contrast,
                                                          r;<than that of

37

_____ liquid hydrogen.[38] Hydrid!;k correlates well with electric batteries found 1n vehicles. Hydrid is the safest storage option.[39]

Gas Under Pressure for Vehicles

Stationary hydrogen stores at low pressures. The reason is the volume of the container is not as crucial as when the fuel is stored in a vehicle. [40] Hydrogen in gas form is the easiest and cheapest method of hydrogen conversion. Its efficiency and range for local travel is around 60 mile. The hydrogen conversion process entails mounting gas cylinders in the truck or back seat of vehicle. Stainless steel fuel lines prevent the absorption of hydrogen gas.[41]

Hythane

Ten percent hydrogen mixes without special modification to the conversion with natural gas. This mixture is Hythane. A 1988 gas under pressure conversion of a postal jeep by a UCLA group cost $580.00. Pressurized gas storage was the form of the modification to the UCLA Jeep . [42]

Liquid Hydrogen Vehicle Comparisons

The 1991, Oldsmobile, a project of the American Hydrogen Association's racing program was converted to liquid hydrogen. The liquid hydrogen tank placed in the car's trunk has a three gallon capacity and provides a range of nearly 100 miles. Compared to gasoline, liquid hydrogen provides three times the energy and weighs less than one-half the weight of gasoline.[43]

Hydrogen Liquefaction Plants

One method of hydrogen liquefaction uses gas pressure produced during electrolysis. Oxygen coming from the electrolyser pressurizes. If allowed to expand rapidly, it absorbs heat from the hydrogen as it passes through the heat exchanger. This prechilled hydrogen needs less energy to liquef y.[44]

Magnetocaloric Effects

Magnetic fields have powerful affects on specific materials, like gadolinium alloys. This material when placed in a magnetic field heats . The material cools when removed. Hydrogen gas cools when it contacts these metals. Magnetic flux transforms develop temperature changes in many materials. The most economical means of liquefying hydrogen might be to use gas compression with magnetocaloric processes.[45] Magnetocaloric effects have mathematical intuitive constructibility.

Ortho to Para Hydrogen Conversion

The electrons in orthohydrogen molecules rotate in the same direction. In parahydrogen the electrons in this molecule spin in opposite directions. When ==-----'liquefied, the orthohydrogen is all converted to parahydrogen. This conversion produces heat which accelerates the evaporation of liquid hydrogen.[46] Ortho to para hydrogen conversions have mathematical intuitive constructibility.

Liquid Hydrogen: Aviation Fuel

NASA and Lockheed have concluded hydrogen is the only fuel capable of propelling hypersonic planes. In contrast, in automobiles liquid hydrogen has both advantages and disadvantages. For example, liquid hydrogen can take 15 minutes to refuel in one converted postal vehicle.[47]

A Hybrid Gasoline Liquid Hydrogen Fuel

At the Ninth World Hydrogen Energy Conference in Paris during 1992, the " F,emch BMW said they were developing a hybrid gasoline-liquid hydrogen. The tank fill is fifteen minutes. Little hydrogen was lost during filling.[48]

"Unlimited Fuel" Class

The manager of the American Hydrogen Association's Racing Program, Demeter Wagner, drag raced an Oldsmobile at 273 mph on 2.3 liters of liquid hydrogen. The .car raced in the "Unlimited Fuel" class. Any fuel such as gasoline, alcohol, hydrogen, nitromethane is in this class.[49]

Automobiles and Hydrides

Hydrides or metals can absorb hydrogen. Automobiles using hydrides have ',,,cit?f·i<.. sl:,ec:lal properties such as rapid absorption, reversibility of reactions, pressure and similarity, rejection of nonhydrogen gasses, high weight holding by ...,,::,,,.•;:""" •,pe,rc,en1ta!11e of hydrogen fuel, durability for 300,000 cycles or more.[50] However, there are many hydrides and dehydrating packing problems. Contamination is less a problem for hydrides using such chemicals as carbon dioxide and oxygen. Different hydrides have different heat exchange systems, for example, magnesium or iron- . titanium hydrides.[51]

Mathematical Intuitionism and Choice

Free choice sequences in hydrogen conversion and in automotive design have intuitive mathematical expressions. For example, engineering .choices form primitive mathematical expressions. Inventions, like experimental cars, have personal choice sequences which define sets. Thus, the whole personalist intuitive framework applies to hydrogen .conversion projects and to invention design.[52]

Experimental Cars and Hydrides

Pontiac 1975

Billings Energy Corporation changed a 1975 Pontiac to hydrogen hydride. The range of the car is about 150 mi. consuming 5.25 lb. of hydrogen. A solenoid valve turns on an off the hydrogen supply power. Performance was not sacrificed in this conversi·on.[53]

Cadillac Seville 1977

To overcome poor mileage and tank weight hydrogen and gasoline were both used in a Cadillac Seville. Hydrogen worked best for short trips. Gasoline was best for lengthier hydrogen trips. In January 1977 President Carter used hydrogen in the inaugural parade in his Cadillac vehicle.[54]

Racetrack Hydride

During May 1984 Kenji Watanabe drove in a hydride powered 1973 Mazda. On the Fuji race track in Gutembe, Japan. He averaged 80mi/hr. It takes between 10 to 15 minutes to refuel the tank. Water injection increases engine power. Future conversion costs would be $1,000 per car in mass prod uction. [55]

A Mercedes Benz Conversion Van

Daimeler Benz develop a van that uses two hydrides: FeTi and magnesium nickel hydride. Hydrogen is removed from the hydride in small cycles. The van also uses hydride air condit ioners. [5 6] Car conversions have mathematical intuitive constructibility.

Activated Carbon Fabric

Roy McAlister head of the American Hydride Association is developing an activated carbon fabric. It stores hydrogen at nearly one third the energy density of gasoline.[57] This is a potentially revolutionary innovation.

Salt Storage of Hydrogen

Salts like sodium formate can store hydrogen. The hydrogen produced is about 30% of the weight of the salts. The change of low value saline fuel within high caloric gas by this means is attractive. In contrast, microstorage of hydrogen into glass spherical containers is also developed by 3-M .[58]

Reforming Hydrogen From Gasoline

Gasoline can reform to hydrogen and other gasses by incomplete combustion utilizing small amounts of oxygen at high temperatures. Using a catalytic reactor, the University of Arizona changed a 40 hp. Volkswagen engine to drive on 100%

_____hydrogen   from
reconstituted gasoline. The car contained just a fuel atomizer for injection and an
engine. Steam and gasoline go to a superheater then move to a reactor where the
mixture burns. The hot hydrogen thermally decomposes the steam creating
hydrogen gas. The condenser then dries the hydrogen gas before the fuel storage.
The precipitation from the condenser recycles.[59] Reforming hydrogen from gasoline
is mathematically intuitively constructible.

Hydrogen comes from Methanol. For example, methane, a form of methanol,
splits into hydrogen and flammable hydrocarbons by engine heat.[60] This is a way
of reforming hydrogen from gasoline.

Hydrogen Fuel: Pros and Cons

Engines changed to hydrogen get only 80% of the power achieved with
gasoline. Hydrogen is 50% more efficient. One form of hydrogen is the gas form.
Gaseous fuels have special advantages: better fuel and air combinations. Fuel does
not precipitate on the walls of an engine that is running in the cold. Gases have cooler
operating temperatures. They give an engine better lubrication. There are few carbon
deposits. What are the disadvantages of gaseous fuels? The faster burning properties
of gases may overheat the valves. Misfiring is a problem. There is lower fuel
economy, a danger of leaks and a danger of overpressurization.[61]

History of Hydrogen Conversion Projects

In the 1870's Otto considered several fuels for his combustion engine, namely
hydrogen. He rejected gasoline because he thought it was much too dangerous.
Gasoline was later adopted. Hydrogen, in early experiences, was mixed with many
gasses such as propane and natural gas. Backfiring was a problem. Water injection
to decrease backfiring made the hydrogen less powerful. During WWI hydrogen
combined with oxygen was a possibility for powering submarines and airships. Rudolf
A. Erres invented the first practical hydrogen engine in the 1920s. He changed 1,000
engines including trucks, buses, and torpedoes to hydrogen. During 1924 Ricardo
systematized engine performance tests for hydrogen. King dealt with hot spots in the
engine. After WWI the University of Miami, with N. R. Swain and R.R. Adt modified
injection techniques with a Toyota engine. The Illinois Institute of Technology used
a propane carburetor to modify a 1972 Vega. Roger Billings and Brigham Young
University won first place in the 1972 carbon vehicle competition with a hydrogen
converted Volkswagen. Robert Zweig survived the backfiring problem by using a
special air intake valve to enter hydrogen from air into a hydrogen powered pickup.
The Brookhaven National Laboratory and Mazda Corporation changed a Wankel or
rotary engine to hydro gen.[62] Hydrogen conversion projects are mathematically
intuitively constructible.

Modern Conversion Experiments

China converted a diesel to hydrogen. Independence, Missouri, converted a postal jeep. U.S. Bureau of Mines, a mining vehicle. [63]

Billings Energy Corporation of India converted a diesel and the

"Fuel Mixing"

There are two types of fuel mixing: external and internal. Both types are mathematically intuitively constructible. In external mixing air and fuel mix outside the combustion chamber. In this approach hydrogen fuel displaces air in the mixture reducing power. Internal air and fuel mixing usually follow a sequence: air injection, closing of the air intake valve, fuel injection, closing of the fuel intake valve, ignition of fuel, and air mixture. There are two types of internal mixture formation: early and later injection. [64]

Uneven mixtures of fuel cause nitrous oxide formation, ignition problems and combustion problems. Increasing the turbulence in the combustion chamber by altering the chamber changes the fuel injection and usually solves the problem. [65]

A Liquid Hydrogen Car

The Musashi Institute of Technology produced in 1975 a special liquid hydrogen car.[66] It is powerful. It is very efficient.

Spark Ignition Engines

The American Hydrogen Association is building a method of changing spark ignition engines to hydrogen power. In contrast diesel engines lack spark plugs. Compression ignites the fuel spont aneously .[67] Ignition is a mathematically intuitively constructible process.

Pollution

Hydrogen produces emission problems through the burning of hydrocarbons from lubricating oil that is burned and nitrous oxide from high temperature nitrogen combustion in the atmosphere. [68] The pollution produced is severe. In Japan, for example, people have to wear gas masks on bad smog days.

Hydrogen and Electricity

Electric cars became public 100 years ago. Electric powered autos are two to three times more efficient or clean than internal combustion engines. However, by 1920 gasoline powered vehicles won over other transportation technologies. The major problem in electric powered vehicles is electrical energy storage in the battery. It cannot rival energy density found in combustible fuels.[69]

42

Hydride Batteries

Hydrogen is the best way to store electricity. During 1987 Ovionics created the first nickel hydride battery. The battery charges by electrolyzing water into hydrogen and hydroxy ions.[70]

Fuel Cells

Fuel cells create an electricity from hydrogen combustion. In a hydrogen oxygen fuel cell hydrogen enters on the anode side while oxygen enters on the cathode side. Water, hydrogen and oxygen exit the cell. Current flows out from the fuel cell. Potassium hydroxide is less corrosive and inputs high cathode voltages.[71]

Solid Electrolytes

The Argonne National Laboratory in Illinois developed a solid electrolyte fuel cell. They used yttria stabilized zirconium made in a ceramic like material.[72] The cells in the solid electrolyte stack like cardboard. The array is a "sandwich" of cells connected in series. Each layer increases the voltage of the next. This voltage is copper on the positive and negative sides of the cell. Electrons released from the anode travel to the cathode of the next cell. The cathode releases oxygen ions to the anode through the electrolyte to repeat the process. The current flows in a zig-zag path to the surface of the fuel stack. Garrett Ceramics Corp intends to mass produce fuel cells utilizing this technology. [73]

Automotive Conversions with Fuel Cells

A fuel cell does not produce torque. It can provide current to an electric motor. A fuel cell has twice the efficiency of a gasoline or diesel fuel engine. A hydrogen powered fuel cell is comparable in cost to premium gasoline. A hybrid fuel cell/battery vehicle meets the performance of regular combustion engine with less pollution. [74] A "reversible" fuel cell called LaserCel 1 while charging electrolyses stored exhaust water to further generate fuel.[75] Starting power generation like fuel cells can adapt to utility power.[76]

On-Site Hydrogen Power Generation

Hydrogen fuel applies in cooking, space and water heating, airconditioners and refrigeration, and lighting and farm implements. In these stationary applications it is superior to synthetic alternatives and conventional fuels.[77]

A Comparison of Hydrogen and Natural Gas

What are the advantages of hydrogen? It has less damaging explosions than natural gas. Leaking hydrogen rises and dissipates faster than propane which has a higher density.[78]

What are the disadvantages of hydrogen? Because of its small size and higher energy kinetic energy it is more likely to leak than gas. Hydrogen can achieve its explosive limit four times as fa st . [79]

Hydrogen is invisible and radiates less heat than natural gas. Because all other gasses excluding helium and neon, liquefy with contact upon liquid hydrogen. Only these gases can purge liquid hydrogen as fuel.[80]

Hydride Power

Hydride discharge of hydrogen can perform work. Solar or waste heat discharges hydrogen gas which turns a turbine. This turbine gives up heat which reabsorbs in a second hydride tank. Heat can then reverse this fuel.[81]

Aphoid Heating

An aphoid heating plant is a plant which burns hydrogen in pure oxygen to produce steam at high temperature. An aphoid heating plant is ten percent the size of fossil-fueled boilers. It has an efficiency of about 90% with few pollutants. The problem is a high combustion temperature required. In contrast an open air burner mixes air with gaseous fuel and inputs a velocity to the fuel to prevent it from burning back into the line again. [82] Aphoid heating is an intuitively constructible process.

Natural Gas Versus Hydrogen

Natural gas is the most common fuel. Hydrogen can adopt to almost every situation where natural gas is an option. Hydrogen flame gives off virtually no visible light. Natural gas burns with a blue flame. Hydrogen is hotter during full combustion 3480 degrees F than natural gas is at 3480 degrees F. Three times the hydrogen as methane is necessary for the equivalent energy. The flow for hydrogen is three times that for natural gas. The flame speed for hydrogen is ten times faster than natural gas. This poses a greater feedback danger for hydrogen. Hydrogen combustion occurs 1n the proximity of certain catalysts like platinum or palladium without a flame.[83]

The model Billings Hydrogen Homestead uses a Tappan Convection range for cooling. For example, the burner of a hydrogen stove heats eight times faster than natural gas. It uses 24 percent less energy.[84]

Safety of Hydrogen

44

What are the safety-related characteristics of hydrogen? First, it has a lower density than air. Thus indoors it must be vented to avoid fire. Its low molecular weight makes hydrogen more prone to leaks than most other gases. Second, it has a higher rate of diffusion in air. In the Hindenburg, hydrogen was not a fuel but stored in balloon-like containers for lift. Many of the 35 passengers who died jumped to their deaths rather than burn to death because the passenger gondola was under the gas cells. Third, the heat energy of hydrogen is less energy than other gases because hydrogen energy on a volume basis is about one-third less energy content than hydrocarbons. Fourth, a 18 percent hydrogen concentration can cause a detonation. In contrast, only 6 percent is necessary for methane and propane. Because it burns quickly, it has the highest explosive potential for mass. Per unit volume, it has the lowest explosive hazard. Fifth, the flammability limits of hydrogen in air are between 4 percent and seventy-five percent. Sixth, the ignition energy (not temperature) for the combustion of hydrogen is only one-third that of hydrocarbons. Seventh, the volume speed of hydrogen is its burning velocity, and the spiral toward the flame displacing an unburnt gas mixture is ten times that of hydrocarbons. Eighth, embrittlement or unwanted hydriding, for example, metal cracks, are a problem with hydrogen. [85] **In short, how can hydrogen be used safely? Three main ways are adequate ventilation, prevention of leaks, elimination of ignition origins.** [86]

"The Technology Trap"

A World Hydrogen Economy Energy dependency has diminished our natural self respect, damaged our environment and wrought war. The Middle East has 75 percent of the earth's known oil deposits. In contrast, the Western Hemisphere has 16 percent. The U.S. and Canada consume twice as much per person as the European countries. America is facing what Roy McAlister calls "the technology trap." For example, in 1999 the U.S. used 6.2 billion barrels of oil. We import 58 percent of it. An alternative fuel is risky and expensive. People want a readily available alternative fuel. The movement is from carbon to hydrogen. The use of coal is declining. However, in a few years coal will replace oil and gas. [87]

Statistics

Subsonic aircraft use 19 percent less energy for liquid hydrogen than fossil fuel. For supersonic aircraft there is a 30 percent advantage. What are the variations in production costs? Petroleum refining accounts for 77% of hydrogen production worldwide? Coal gassification accounts for 18% of hydrogen production worldwide. Electrolysis accounts for 4 percent hydrogen production worldwide. Transmitting hydrogen by pipeline is one and one half as costly as natural gas. The environmental damage of petroleum is $1,613 billion or one-eighth of the worlds volume of goods and services.[88]

45

The key to hydrogen use is the refinement of water electrolysis, a mathematically intuitively constructible process. [1] Professor Doyle Britain claims the electrolysis of sea water is not a difficult <u>chemistry</u> problem. [2] This electrolysis of seawater is the Neptune Project named after the Italian god of the sea. Electrolysis of sea water is a special issue. Water covers seventy percent of the earth's surface. This amount of water will precipitate the use of hydrogen in vehicle propulsion. Not all areas of the earth have ready access to water, for example the desert areas. So gasoline and electric vehicles will have a definite role in these areas.

What are the major approaches to sea water distillation? There are presently four major types of sea water distillation: vapor-compression distillation, reverse osmosis, flash distillation and electrodialysis. Each of these types as the following chart shows has different energy consumption, water quality and recovery rates.

Energy Consumption

Energy Consumption MJ/m3 H20		Water quality,ppm	Recovery
vapor-compression- distillation	86	5	57%
reverse osmosis	19	150	50%
flash distillation	133	5	25%
electrodialysis	38	500	3 3 % [3]

The keys to the electrolysis of sea water are energy consumption and recovery. A third key is water quality. The salt by-product of desalination, however, has many uses, especially in food preservation. In general the whole mathematical framework of intuitive constructibility analyzes the various methods of electrolysis minutely.

The Revision of History and the End of Colonization

The successful use of sea water for hydrogen conversion will free many countries, like Israel, Ireland and African countries, from external colonization. In addition, Germany and Latvia will have a replenishable energy source. If history means empires, hydrogen technology may bring about the revision of history as we presently know it. The movie, <u>Chain Reaction</u>, highlights the powerful forces hydrogen will uncap for the world. It will bring about the death of most present day empires.

HYDROGEN VERSUS ELECTRIC VEHICLES

James Mackenzie finds Japan will have 200,000 electric vehicles in production by 2000 and will build 100,000 of them yearly from then on. In 1995 France will begin mass production of electric vehicles to stimulate the move to electric vehicles. Mackenzie recommends (1) raising fuel prices to make electric vehicles more cost competitive with gasoline-powered vehicles, (2) more shared cost research and development on battery and hydrogen technology (3) infrastructure development like recharging batteries, hydrogen alternatives, and battery-recycling centers, (4) market stimulation and educational incent ives. [1]

During the past 100 years, petroleum-powered vehicles have revitalized American life. Mackenzie allows the progress for electric vehicles is brighter than the three major carbon-based fuels. EVs powered by batteries, hydrogen based fuels, ultracapacitors and fly-wheels are the most promising. [2]

Oil-powered vehicles have many negative-social changes such as air pollution, acid rain, greenhouse gas emission, dependence on foreign oil resources and a large trade deficit. In absolute terms, net oil imports rose by nearly 60 percent between 1983 and 1992. Net imports are U.S. imports minus U.S. exports. Persian Gulf oil imports increased from 4 percent of the supply to over 12 percent between 1983 and 1992. The National Energy Strategy finds the U.S. bill will increase by almost 6. 7% per year or from $60 billion in 1990 to 200 billion in 2010. [3]

Carbon dioxide is the most significant greenhouse gas and accounts for 67 and 96 percent of the earth's future warming. The increase from 74 million motor vehicles in 1974 to 191 million in 1991 has sandbagged all progress in carbon dioxide emissions. The transportation sector accounts for one third of carbon deposit emissions. [4]

Car efficiency improved from 13 mpg in 1973 to nearly 21 mph in 1990. Motor vehicle use increased by more than 40 percent during this time while oil imports doubled. The size of light trucks increased to 36 percent of all light-duty vehicles. [5]

By 1995, 10 percent of all new vehicles must use alternative fuels. Zero Emission Vehicles must be two percent of the new cars by 1998. [6] This is true is the United States but more true in select countries abroad, like Japan and France.

What are the alternatives in vehicle technologies? First, Methanol or wood alcohol comes from natural gas. However, Methane, a derivative, releases the same amount of greenhouse gases as gasoline vehicles. [7] Second, Compressed Natural Gas is used in Italy and New Zealand in motor fuel. The problem with it is it is bulky and refueling problems are difficult. Compressed natural gas vehicles reduce

47

greenhouse gas emissions for light vehicles by 15 percent.[8] Third, Ethanol, a grain alcohol, is part of alcoholic beverages. Gasohol contains nine parts gasoline to one part ethanol accounts from nearly 6 percent U.S. nation fuel use. [9] Ninety-five percent of ethanol used in motor vehicles comes from corn. Gasoline would increase ozone concentration by 6 percent.[10] By enlarge ethanol, methanol and natural gas have few significant benefits over gasoline. [11]

Electric Vehicle Options

Batteries, flywheels, ultra capacitors, and hydrogen fuel-cells are technologies to power electric vehicles. They are reliable, need little maintenance and are virtually pollution-free. Their greenhouse gas emissions are slow. The problem with electric vehicles it that they are more costly, have shorter ranges and take 15 minutes to 8 hours to recharge.[12]

Japan and Germany are developing hydrogen fuel-cells with rotary engine internal combustion engines. The fuel cell hydrogen powered vehicle has twice the range of hydrogen vehicles powered by an internal combustion engine. Most hydrogen today comes from natural gas an unrenewable energy resource. Electrolysis, photochemical reactions, renewable or nuclear technologies are other sources. [13]

History of Electric Powered Vehicles

Robert Davidson, a Scottish inventor, in 1833, built the first prototype battery powered vehicle. America saw its first commercially produced battery powered vehicle come out in 1894. It was called the Electrobat; and, it was produced by Morris and Salom Company of Philadelphia. By the late 1890s, EVs had a fifty mile range. By 1900, 38 percent of new American automobiles ran on batteries. Before WWI, one third of motor vehicles in America were electric and often powered by Thomas Edison's nickel-iron battery. Early EVs had a twenty-five to 40 mile range. The electric starter gave the advantage to gasoline-powered vehicles. In 1982 Denver introduced 40 foot battery-powered buses. GM is targeting mass production of EVs from the late 19 90 s. [14] In 1977, Japan had 13,000 electric vehicles and England 30,000 electric vehicles in operation. [15]

Recent Progress

In 1988 General Motors, with inventor Paul MacCready of AeroVironment Inc., released the electric sports car, the Impact. It can go from Oto 60 mph in 8 seconds, has a top speed of 75 mph and a range of 75 mph .[16] From 1982 to 1985 Ford developed the ETX-I and electric version of the Ford Escort. In 1990, the company developed and electric version of its Aerostar van, the ETX-I1. It had a range of 100 miles, a 65 mph top speed and a 1,000 payload capacity. In 1993 Chrysler developed

50 electric vehicles, the Dodge Caravan Electric. They had a range of 80 to 120 miles, a 65 mph top speed, and accelerates from 0to 30 mph in 8.25 seconds.[17]

Most other small U.S. companies developing EVs are converting commercially available internal-combustion engines to EVs. There are many of these companies. Also, the Electric Transportation Coalition and the Electric Vehicle Association of the Americas are just two of the EVs promoting organizations in the United States. The Department of Energy is the main source of federal funds and developm ent .[18]

Japan is hoping to release 200,000 electric vehicles by the year 2000. In Europe the European Electric Road Vehicle Association promotes electric and hybrid vehicles. Citelec and other organizations of European cities combine to produce EVs in urban are as. [19]

Citroen in France is building a City Electric Automobile using nickel-cadmium batteries which achieve a top speed of 68 mph, accelerating to 31 mph in 8.5 and have a city range of 68 miles. Renault also introduced two electric vans in 1993. Paris is introducing an EV fleet in 1995.[20] Most EVs use lead-acid batt eries. [21]

The Intuitionist Framework, Electricity and Electric Cars

In general the whole intuitionist framework of interpretation and analysis applies to electricity, electric cars and electric inventions, like ultracapacitors. The electrons in electricity are a constructible activity. Intuitive free-choice sequences are integral to electric cars design and to electric inventions, like batteries. Electric cars are personal constructible activities. [22]

Ultracapacitors and Flywheels or Elector-Mechanical Batteries

The main alternative to Batteries are ultracapacitors and flywheels An ultracapacitor is an electric component that canstore large amounts of electric energy. Flywheels outperform batteries in every way. What is a flywheel? It is a rotating object that possesses kinetic energy which can be converted to electrical energy. Magnets placed on the rotating wheel develop power as their magnetic fields cross wires around the wheel. The process reverses making the flywheel rotate faster and storing much more energy .[23] An electro-mechanical battery, as flywheels are termed, has specific energy demands as high as 150 Wh/kg. Pound for pound, flywheels pack nearly ten times as much power as a standard V-8 engine and are seventy times as powerful as a lead-acid battery. The rotor of an EMB spins on magnetic bearings in a vacuum at nearly 200,000 RPM. They have almost no friction and self discharge over long periods of time. They recharge easily and store electric power and regenerate easily. Honeywell, a local Minneapolis company, is one of many building flywheel systems in Americ a. [24]

_____Hybrid

Electric Vehicles

Hybrid electric vehicles that supplement electric batteries from other energy sources such as gasoline, fuel cells, ultracapacitors and flywheels are also being developed. The hybrids should average out problems developed with other electric developments. Los Angeles is plugging the electric vehicles in a formidable way. Life cycle cost estimates vary quite radically. [25]

The best way to use hydrogen to move vehicles is through fuel cells. The wheels of EVs containing these cells would power by electric motors which emit no air pollution . [26] Fuel cells are a win-win situation.

The Many Types of Fuel Cells

A fuel cell, like a battery, has no parts that move and changes chemical energy into electricity efficiently. Hydrogen-oxygen fuel cells get efficiencies from 50 to 60 percent.[27] There are several types of fuel cells. Phosphoric acid fuel cells heat and provide electricity in malls. Their size and weight make them used best in heavy duty vehicles. Alkaline fuel cells are part of space flights. Solid oxide fuel cells have longer term staying power for vehicles performance. Proton exchange membrane fuel cells allow for rapid start-ups at normal outside temperatures. [28]

Findings

How would zero emission vehicles affect ecology and the economy? Many zero emission vehicles will radically drop oil consumption, greenhouse gas emissions and air pollution .[29] Electric utilities would be dramatically affected by a growth in the number of EVs. An electric vehicle might increase a family's electric use by 22 to 55 percent. Nighttime recharging would reduce smog fo rmat ion .[30] Also, it is much cheaper to produce electrolytic hydrogen and oxygen during off-peak hours;[31] and, catalysis is the major problem "for the direct storage of solar energy in photoelectrochemical cells."[32]

Hydrogen and Electricity: The Opposite Sides of the Same Coin

Hydrogen-Powered EVs

Hydrogen and electricity are opposite sides of the same coin. Electricity can produce hydrogen through electrolysis. On the other side of the coin, hydrogen can be consumed to create pollution-free electricity through a fuel cell which unites hydrogen and oxygen chemically to create electricity, water and waste heat. These battery-like devices use no high-temperature combustion and develop none of the nitrogen oxides necessary to form smog.[33] Thus, the intuitive constructible activities of hydrogen and electricity form the opposite sides of the same equations.

A Hydrogen "City of the Future"

The political option, a personal intuitive free-choice sequence, for hydrogen urban development is a national priority. A "City of the Future" is in the planning st age. [34] Sustainable growth in the next century in American cities will base itself on hydrogen and electricity and on hydrogen as an energy carrier.[35]

51

The purification and sanctification of space and time is an issue in hydrogen research and project s.[1] The idea of the common good is one way to do this research and these projects. Hydrogen is a symbol of purity which is a key way to sanctify the universe. It is a pure fuel. What is the common good? It relates to personalism. So does the use of hydrogen. Fuel is intuitive, and its use has a personal human dimension.

THE COMMON GOOD

John **XXIII** in the Encyclical <u>Pacem in Terris</u> suggested the common good is best protected when a person's rights and duties are guaranteed. Civil authorities must assure these rights "are recognized, respected, coordinated, defended and promoted, and that each individual is enabled to perform his duties more easily . " [2] Rights are inviolable. If a government does not recognize human rights, its decrees will not be bindin g.[3] The common good defines life rights. Why is there not agreement on a common God? Should not a definition of the common good relate to the idea of a common God? The answer is there is only one true God, the Trinity, because it has a social dimension for eternity. Wars over religion are the rule rather than the exception. The common good is the fundamental basis of the just war and of all levels of culture, including the culture of ecology and hydrogen use.

Solidarity in the Rights of Life and The Common Good

Solidarity: is the "combination or agreement of all elements or individuals, as of a group; complete unity, as of opinion, purpose, interest, feeling, etc." and culture.• Solidarity, for example, is social and political and is a way of defining rights and the common good. Life must further the common good.[5] T.S. Eliot thought unity, like solidarity, is the basis of genius while conformity is the mark of mediocrit y.[6] One earth, one people are the best forms of solidarity. Hydrogen use develops best through solidarity and subsidiarity.

Authority

Authority comes from God so civil law must conform to the moral law. **It is never legal to help in evil.**[7] The Church should protect the faithful from politicians, wayward intellectuals and mad scient ist s. [8] My father, Aquinas O'Dougherty, quoting St. Thomas Aquinas says "where there is a choice between two evils. There is no choice." This is often the case in politics. However, there **is** a choice between good and evil.[9] One of the strengths of the Catholic Church is that it is not a democracy. Its appointments are not based on political whims. It looks at the "test of time." Many politicians just want your money and your life. For example, the Vatican has condemned the American legal system for being morally bankrupt and corrupt and perhaps rotten. Something is rotten in America--the American earth. Ecological and scientific authority also come from God. Perhaps the reason the earth has become rotten is because of corrupted and rotten earthly authority. Hydrogen use can purify the rotting earth.

Ecological World Parliament

To go into effect, the hydrogen conversion project calls for a worldwide formation of a green Federation of the Earth Party to ratify the Constitution for the Federation of the Earth and to deliver delegates to the World Parliament. [10] This constitution and this world parliament should back nonprofit environment organizations. They should grow from charity contributions.

Solidarity and subsidiarity should be the principles of the union and integration of this Constitution for the Federation of the Earth and of the World Parliament of the ecology movement. Solidarity has been discussed. What is subsidiarity? First, it is a principle of union and integration. For example, it is the anchoring of hydrogen and electric vehicles research within environmental protection. Second, it is the environmental compatibility of all energy conversion proposals. Third, it is a social agenda as parts of any energy conversion. Fourth, it is dialogue as the basis of the relation of social, political, and governmental partners in the integration process for the earth's economy. Fifth, it is integration that differentiates, for example the differential impact a hydrogen economy would have on local communities throughout the world. Integration should remain permeable and emphasize transitional arrangements. Objective criteria should determine participation in integration. Sharp class divisions run counter to integration of hydrogen. [11] Thus solidarity and subsidiarity are the keys to the development of hydrogen and to the development of the Constitution for the Federation of the Earth and a World Parliament.

53

A Definition of Right or Rights from Father Hilary Freeman

Father Hilary Freeman suggests there is "right in the juridical order, what corresponds to juridical order, and powers belonging to members of a   juridically ordered community. "[1] He contends "only persons can be the subjects of rights and duties. The virtue of justice has to do with legal obligations and the rights of others. "[2] A freeman has juridical rights, which correspond to juridical order,   and powers belonging to the juridical ordered community. In contrast, is a slave defined as a person or property? The answer is in many countries a slave was the property of the slaveholder. Materialism denies freedom through determinism .[3] Relationships must change in nature from materialistic relationships to will bonded relationships. Personal rights and duties must be the framework used for the development of a hydrogen economy for all peoples. People should have energy rights and duties.

A Catholic personalist intuitionist rights and responsibilities approach to ecology would sanctify space and time through the common good. How would this happen? There is a Catholic rights movement in the Church which applies to ecology.

Catholic Rights in the Church: A Charter
Prelude
What is the charter of this Catholic rights movement? Catholic rights and responsibilities, for example, derive from their humanity, baptism, and membership in the Church. Rights guarantee the dignity and freedom of the person and the Catholic. The United Nations Charter stated basic human rights. The Church Q_@Supplements these rights with those given at baptism supposedly (1) the priesthood of believers, (2) the equality of believers, and (3) the prophecies of believers.•

The sponsors of the Charter of Rights for Catholics follow the Gospel message of justice within the Church. For example, the Church as a People of God not as individual Christians should give witness to the commandments. This witness creates a responsibility to renew the Church's structure to create justice. Christ destroyed all divisions such as race, nationality, sex, age, social position and states-of-life. This charter maintains Catholics are radically equal. Rights exist with responsibilities. There exists a loving relationship between the magisterium and the People of God.[5]

Fundamental Rights

1. Catholics have the right to follow their educated consciences in all issues.

2. Church officials should turn to the laity for consultation on issues of private and public morality.

3. Catholics have the right to participate in activities which do not infringe upon the rights of others, for example, the rights found in the Bill of Rights such as freedom of speech, press, and association.

4. Catholics have the right to obtain information held by Church officials on their spiritual and temporal well being, providing such access does not impair the rights of others.

Decision-Making and Dissent Rights

5. Catholics have a right to a voice in all decisions that concern them, including leadership selection.

6. Catholics have the right to expect accountability from their leaders.

7. Catholics have the right to make voluntary associations to develop Catholic aims.

8. Catholics have the right to public dissent about decisions made by authorities, including Church authorities.

Due Process Rights

9. Catholics have the right to quick and fair judicial procedures.

10. Catholics have the right to regular legal procedures for redress of grievances.

11. Catholics have the right to privacy and protection against character assassination.

Ministries and Spirituality Rights

12. Catholics have the right to Catholic instruction, worship and pastoral counseling.

13. Catholics have the right to follow their unique callings in life, and both personal and community guidance.

14. Catholics have the right to follow the customs and rites of worship of their choice.

15. Catholics have the right, regardless of race, age, sex, nationality, state-of-life and

social position to receive the sacraments if they are correctly ready to do so.

16.   Catholics, either lay or clerical, male or female, have the right to exercise ministries in the Church commensurate with the approval of the community and meeting the standards of Church preparation.

17.   Catholics have the right to expect a sense of community from Church officials.

18.   Church officials have the right to training, fair pay, and freedom in the exercise of their positions.

19.   Catholics have the right to expect proper training and education from Church officials.

20.   Catholic theologians have the right to academic freedom, dialogue, and pluralism of belief.

Cultural and Social Rights

21.   Catholics have the right to political freedom.

22.   Catholics have the right to follow their informed consciences while pursuing justice and peace.

23.   Church employees have the right to just wages and decent working conditions.

24.   Catholics have the right to artistic and cultural freedom.

Rights of States of Life

25.   Catholics have the right to select their state in life: single, married, or celibacy

26.   Catholic women have equal rights with Catholic men.

27.   Catholics have the right to expect fair distribution of Church resources, on education, the single state in life, without regard to race, nationality, sex, social position, and sexual orientation.

28.   Catholics have the right to determine the size of their families.

29.   Catholic parents have the right to see to their children's education.

30.   Catholics have the right to divorce and remarry. The Association for the Rights of Catholics in the Church claims Catholics have this right but the Church has never approved this claim.

31.   Divorced and remarried Catholics have the right to Church ministry and the sacraments. ARCC maintains Catholics have this right but the Church has never sanctioned this claim.

32. Church documents and materials should not be sexist or exclusively masculine. [6]

Sexuality Rights

To counter Fascist sex experiments and sex atrocities, all of these rights and responsibilities issues must converge in a "seamless garment" of unity. For example, this "seamless garment" ethic should counter the psychopathic, schizophrenic Nazi sex experiments. In contrast, the ethic of life and sexuality and its definition in literature and ecology are a "seamless garment" for all races and all peoples and for all levels of ecology, earth constitutions, the world parliam ent .[7] The life issues and sexual issues create the universal basis of a literature and film of life and life themes and ecology themes, for example, pro-life all species. The snakes St. Patrick drove out of Ireland were sexual snakes.

The Military and life and the Rights and Responsibilities Issues

Jeanne Holm's, Women in the Military: An Unfinished Revolution makes the case for another American Revolution. [8] The foundation of the Green revolution should be rights and responsibilities of women and minorities in the military and also within and without the military context throughout the world. Moreover, the third or fourth most powerful nuclear force in the world is North Dakota. The human species has a right to life.

Personalism, the Catholic Rights Movement and the Ecology Movement

Personalism is a philosophy based on the rights and centrality of the individual. Ecology is personal and intuitive. looking at the charter of Catholic rights, how is a

Catholic rights dimension applicable to the ecology movement? In ecology persons and issues have a dignity, freedom and justice dimension based on fundamental rights like conscience, freedom of speech, association and information rights. In ecology persons and issues have decision-making and dissent rights, like leadership selection, accountability, association right and public dissent rights. In ecology persons have due process rights, for example rapid legal procedures, redress of grievances, and right to privacy. Ecology is a spiritual movement with ministries and spirituality rights such as worship, education and counseling, rites of custom, sacramentality rights, right to a sense of community, right to education, training and fair pay in ecology positions, right to dialogue and pluralism of belief. People involved in ecology have cultural and social rights such as political freedom, decent wages and working conditions, and artistic and cultural freedom. People involved in ecology have the right to select their state in life, for example, single, married or clergy. Each of these states has different rights and responsibilities. Men and women have equal rights in ecology. The right to determine the size of a family is an ecology right. Education for children involved

57

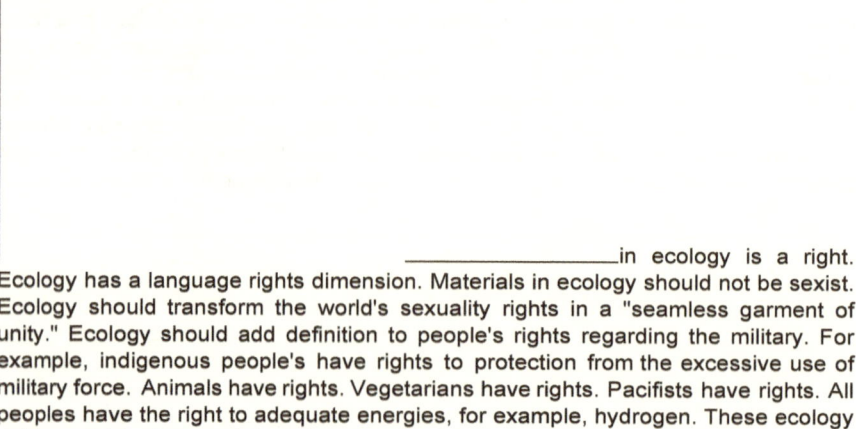

_____in ecology is a right. Ecology has a language rights dimension. Materials in ecology should not be sexist. Ecology should transform the world's sexuality rights in a "seamless garment of unity." Ecology should add definition to people's rights regarding the military. For example, indigenous people's have rights to protection from the excessive use of military force. Animals have rights. Vegetarians have rights. Pacifists have rights. All peoples have the right to adequate energies, for example, hydrogen. These ecology rights exist in the analytical framework of intuitionism and in the context of the common good.

A Critique of the Rights Approach from Mahatma Ghandi

Mahatma Ghandi thought "a society driven by responsibilities is orientated toward service acknowledging other points of view, compromise and progress-- whereas a society driven by rights is orientated toward acquisition, confrontation and advocacy. " [9] In contrast, Martin Luther King favored the rights approach. Both approaches have strengths. Moreover, for each right a person has a corresponding responsibility or responsibilities. For example, in America people have a right to free speech. They also have a corresponding responsibility to avoid character assassination which was mentioned in the above section on due process rights. Also, all species have an ecological right to life. This position entails numerous common good responsibilities.

Christ is the Green in the Ecology Movement

All species have a right to life. We are all in this together. Like St. Patrick and Richard Feynman, we should try to transmit all relationships in nature by sending life to all of the elements throughout the universe. As was mentioned earlier, we should try, like St. Patrick, to make "the inhabitable places habitable. Science fictions writers call this terra forma. "[10] St. Patrick, Christ, grass and frogs are just some of the greens in the ecology movement.

58

THE INTERNATIONAL ALLIANCE FOR SUSTAINABLE AGRICULTURE: NEWMAN CENTER--A CASE STUDY--A NEW CYCLE

The International Alliance for Sustainable Agriculture is a spinoff of the International Federation of Organic Agriculture Movements. Its focus is sustainable agriculture projects, farming and research around the world. The Minnesota Chapter of !ASA is in the Dorothy Day Room at Newman Center at the University of Minnesota. Like the writer, Dorothy Day was a personalist. !ASA "is a nonprofit, tax exempt organization of individuals and groups cooperating to develop ecologically sound, economically viable, socially just and humane agriculture systems around the world. Its three major program areas are: 1) Organizational Support, Network-Building and Collaboration; 2) Education and Information Dissemination; 3) Policy Development. IASA is funded by memberships, the MN Environmental Fund, Clare B. Hilliker Fund, Skiers Ending Hunger, sale of publications, donations, honoraria, corporations, religious institutions and foundations. "[1]

At Newman Center at the Minnesota Chapter, Terry Gips is the co-founder and president emeritus of IASA. Patrick A. O'Dougherty is an activist, a founder of the personalist intuitionist school of philosophy and physics, and founder of the Hydrogen "Conversion" Project. Leo Cashman is heading a Dental Mercury Awareness Project. Kathy Haskins is one of many members and activists.

Parallels to Richard Feynman and St. Patrick: Terry Gips and Patrick O'Dougherty

Terry Gips, Ph.D.: A Profile

As IASA starts its second decade, Terry Gips has spent the last 10 years trying to save the earth from eminent destruction. With this idea in mind he joined the Third World Institute together with several farmers, researchers and consumers to co-found the IASA in 1983. At that date, sustainable agriculture was considered by many to be unworkable. Ten years later the sustainable agriculture movement and IASA have spread worldwide with the sustainable agriculture goal accepted by 170 countries at the U.N. Earth Summit held in Rio. Terry Gips worked at a low level salary based on his concept of just livelihood and Gandhi's idea '"there is enough for everyone's need but not for their greed. " [2]

During the last ten years Terry has be able to

(1)   ) develop a widely-accepted definition of sustainable agriculture;
(2)   write the celebrated <u>Breaking The Pesticide Habit</u>:
(3)   co-edit <u>The Humane Consumer and Product Guide:</u>
(4)   document sustainable agriculture in 45 plus countries and produce audiovisual presentations such as "A Grain of Hope" and "Toward a New Agriculture;"
(5)   co-edit <u>Manna</u> and <u>New Directions in Agriculture</u> and write chapters in <u>The Mother Earth Handbook, The 1992 Earth Journal</u> and <u>Planting the Future</u>;

(6) serve on many boards including the International Federation of Organic Agriculture Movements, the National Coalition Against the Misuse of Pesticides, CERES, Pesticide Action Network of North America, and an advisor to Give to the Earth Foundation;
(7) worked to found the Humane Sustainable Agriculture Project to insure, for example, the ethical treatment of animals;
(8) led the Sustainable Agriculture Policy Group to enact critical legislation to ban pesticides and back sustainable agriculture;
(9) worked to have Aveda Corporation sign the CERES Principles which favor corporate environmental responsibility;
(10) in Minnesota he served on the Minnesota Pollution Control Agency Waste Hazardous Pesticide Task Force, and the MN State Ground Water Environment Quality Board.
(11) gave many talks and media interviews;
(12) raised money for Skiers Ending Hunger;
(13) served as an official representative in a non-government capacity to the Preparatory Committee meetings and for the June 1992 U.N. Earth Summit in Rio .[3]
(14) served as a devout Jew at many synagogues.
(15)) Terry Gips definition of "sustainable agriculture" contains the dimensions of ecological soundness, economic viability and social justice. [4]

60

_____Patrick A. O'Dougherty, Ph.D.: Irish White Negro--A
Profile

Patrick A. O'Dougherty, Irish White Negro, activist, writer, and intellectual has been with IASA for 6 1\2 years building the resource center library and data base. Since his childhood, he had a dream like Martin Luther King to brace the non-white experience in the world with that of the Irish White Negro. He chose the writing constructs of Catholic, wanderer and scientific revolutionist for definition. What are some of his contributions?

All the Tough Numbers
Maximum Sacrifice for the Common Good

Patrick O'Dougherty has done all the tough numbers: published in grade school, attended St. John's University which is a monastery school, academic suicide in ROTC, a seminal member at the University of Minnesota in the Students for a Democratic Society and the Young Socialist Alliance, graduate school at the University of Minnesota, running for mayor of Minneapolis in the DFL caucus, psychology experiences including psychoanalysis, the deciding effort with Hubert Humphrey in the Henry Kissinger appointment, existentialism and radical behaviorism, eight months on Bellwood Chain Gang in Atlanta in the image of the slave St. Patrick, exile, research at Widener Library at Harvard, a trip to Northern Ireland during the Troubles, businesses in the Black community, a Benedictine Oblate for many years, involvement with the Harvard/Radcliffe Benedictines, the Legion of Mary for eleven years, seven books including a published Ph.D. in the history of science, a Royalist/Radical Publishing Company, teaching in the Trio Program in General College at the University of Minnesota, the building with a few others of a world class agriculture research library at the International Alliance for Sustainable Agriculture at the Newman Center at the University of Minnesota, a member of the Austrian Studies Department for eleven years, a wanderer as methodology, inventor of the personalist intuitionist school of philosophy, single and celibate, a Catholic/Jewish dialogue at Temple Israel for four years, the founding of a movement to stop female circumcision in the world, the project to convert the ICBMs in the Dakotas to a Northern Space Center, the effort to establish a Nobel Prize category for women's contributions, the try for Nobel Prizes, the project to stop the holocaust of the children of America (PHCA), the project to convert the world's economy to hydrogen fuel, the founding of the Minnesota Catholic Academy of the Sciences, a founding activist in the alternatives to petrochemicals movement which is the most formidable research area in many fields, for example, medicine. Finally, Patrick A. O'Dougherty, Ph.D., is a wanderer and the founder of the Green, the Grey, and the Purple, Viconian Revolutions. Vice was the father of the social sciences.

A List of Publications: Patrick A. O'Dougherty, Ph.D.

(1)   Patrick's "Unfinished"

(2) An Existential Approach to American History

(3) Walden III: **A Catholic America**

(4) Reinventing Physics: Logic and Physics

(5) Shaking Up Shakespeare: His Dream Work and Personality

(6) "The Classical Father": **A** Children's Story

(7) "A Critique of the Catechism of the Catholic Church"

(8) "A Critique of Newman Center"

(9) Personalism and Mathematics as Women's Personifestoes

      This last work personalizes the intuitionist school of mathematics.

(10) Life Culture Versus Death Culture and the Death of Issue

Agriculture Research Library

The major research library in sustainable agriculture in the world was built for the International Alliance for Sustainable Agriculture located in the Dorothy Day Room at Newman Center at the University of Minnesota largely by Dr. Patrick A, O'Dougherty. It is the Max Planck Library of low capital input agriculture research.

ACTIVITIST AND THEORETICIAN IN THE ALTERNATIVES TO PETROCHEMICALS MOVEMENTS

62

A Flyer from the HII
Hellenist International Institute
A Royalist/Radical Publishing Company
"Be Ye Perfect, Therefore, As My Heavenly Father is Perfect."
Matthew 5:48
PROJECT NOBEL PRIZES: ALL CATEGORIES
PROJECT HYDROGEN CONVERSION
Remember the Hindenburg Blimp!
A PATRICK O'DOUGHERTY PROJECT TO CONVERT THE WORLD'S ECONOMY TO
HYDROGEN FUEL

Hydrogen fuel is the most common, most efficient, most powerful and most clean fuel. It can power automobiles, planes, even space missions. The reason we don't use hydrogen in automobiles is because of the Hindenburg Blimp crash. The claim was made that it was too unstable as a fuel. It is more stable than fossil fuels. Most countries, for example, Israel, China and Poland do not have many fossil fuels. Solar panels are too inefficient. The International Alliance for Sustainable Agriculture is building research and library building projects on hydrogen conversion world wide. The world should not place all of its eggs in one basket. It took millions of years to build up the fossil fuel reserves in nature. It would be a tragic mistake to deplete these reserves rapidly. War would be the outcome. In Japan people have to wear gas masks to protect themselves from the pollution. In this country Lake Erie is dead. The byproduct of hydrogen is water vapor which could be used to irrigate. The primary economic pro"blem in most North American states is fuel and gas importation. This is true for Minnesota. A spinoff of this project would be a hydrogen powered Northern Space Center.

A Personalist Intuitionist Philosophy at Newman Center and IASA
A Postcolonialist Philosophy
IASA is housed in the Dorothy Day Room at Newman Center. Dorothy Day was

a personalist. I'm an intuitionist personalist. Intuitionism is the activity school of mathematics founded by Jan Brouwer. Intuitionism is the idea that "mathematics is an activity," as opposed to "Plato's ideals." See Michael Dummett, <u>Frege: Philosophy of Mathematics</u> (Cambridge: Harvard University Press, 1971). The research projects at Newman Center and at IASA area activities. Mathematics is one magnitude of these activities.

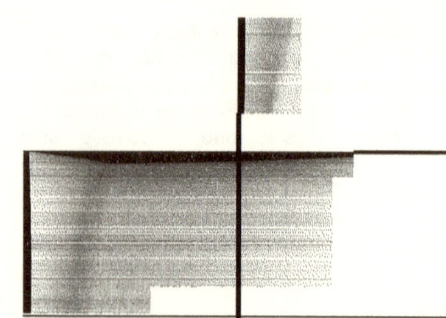

FOUNDING OF THE INTUITIONIST SCHOOL OF PHYSICS BY PATRICK

QglQIJQIiEFITYAUGUST1995

The IASA AGRICULTURE LIBRARY

The IASA Agriculture Research Library is largely a project of Patrick O'Dougherty, Ph.D.

A PROJECT TO REFOUND WORLD CULTURE AT ALL LEVELS

Riverside Plaza M3410 1615 So. 4th. St. Mpls. MN 55454
(612) 339-1748

THE ECOZOIC

AGE

Progress

Progress has been gauged not by the florescence and integral functioning of the Earth's habitat and community but by human control over the nonhuman world for the benefit of humans. For example, the rising Gross Domestic Product contrasts with the developing Gross Earth Product. This is the flaw in the utopia myth.[1]

A New Age: Ecozoic

Swimme identifies a fourth biological era the Ecozoic era. It succeeds the Paleozoic, the Mesozoic and the Cenozoic. These last three eras are mythical expressions that originated in the nineteenth century to help us think about the greater systems of functioning of the planet's ecosystems. They are subjective expressions based in the objective world and not found in the observable earth. The Ecozoic era is an emerging era.[2]

Communion: The Central Premise of the Ecozoic

Swimme claims the universe is a communion of subjects rather than a collection of objects. This is the central premise of the Ecozoic. Intimacy is found in the interrelatedness of all beings. Organic which is a communion develops the inner coherence and integral functioning of the earth. It corresponds to the interrelatedness of the members. [3]

Ecozoic Themes

What are the major themes of the Ecozoic Age? First, "The well-being of the Earth is primary. Human well-being is derivative." Second, balance is the key to the evolutionary process. Third, the intimacy of relatedness corresponds to the seasons, menstrual cycles, and to the universe. Fourth, the Ecozoic is about the return to "virgintime" on the earth. Fifth, from cycles we move to a sense of the irreversible transformations of nature. Sixth, "The earth is a one-time endowment." Seventh, to contrast the ecozoic era, Swimme uses the term the technozoic era which is the controlled order of things that the larger part of industrial society orient towards. [4]

An Ecozoic Dictionary: A New Language for the Earth

Language is destiny. For example, growth has limits. And well people cannot inhabit a sick earth. Biocide and geocide are evils that go beyond suicide, murder and genocide. We need a new Ecozoic dictionary which relates ideas like growth, sickness, biocide and geocide, society, good and evil, freedom and progress to the natural world.[5]

The TransPersonalist/Intuitionist Ecozoic Age: A Communion

The Triumph of Earth Day Over May Day

The concept which best relates the transpersonal, the personal, intuitionism, the ecozoic era to the universe is communion. This concept forms the basis of thought of a priest St. Patrick who rose up from slavery and relates to the thinking of Richard Feynman, a Jew, who wants to change all the atoms within the communion of the earth. All of the writers and themes in this book harmonically converge in a communion of the shared faiths of the ecological movements. David Noble thinks physics is basically theology. So is ecology. A personal spiritual intuitive transfusion or communion to all the matter in the universe, especially the dominant element hydrogen, during the ecozoic era is the best willed analytic approach. This approach should result in the ultimate triumph of Earth Day over Marxist/Socialist May Day in a Green Everevolution.

66

THE INTERNATIONAL
ALLIANCE FOR SUSTAINABLE AGRICULTURE
PURPLE DATA BASE[1]

"The color purple" was an aristocratic choice of color in Europe. This is the reason the writer calls it the purple data base. Ironically, it is also the name of book by Alice Walker on abused Black w omen.[2]

This is the IASA Library and Data Base. It is largely the work of Terry Gips, Patrick O'Dougherty and Jean Carruthers. Patrick O'Dougherty has put six and one half years work into this library. The following list was largely created by a student intern at IASA, Todd Wetzel, and by Terry Gips, Patrick O'Dougherty and Jean Carruthers. Most of the files have more and one entry and many have a great deal of entries.

Country Files

Africa
Argentina
Asia
Australia
Bangladesh
Belgium
Belize
Bolivia
Botswana
Brazil
Burkina Faso
Canada
Cape Verde Island
Caribbean
Chile
China
Columbia
Costa Rica
Cuba
Czechoslovakia
Denmark
Ecuador

Egypt
El Salvador
England, Scotland, Wales
Ethiopia
Europe
Finland
France
Germany

67

Ghana
Guatemala
Haiti
Honduras
Hong Kong
Hungary
India
Indonesia
Ireland
Israel

Italy
Jamaica
Japan
Jordan
Kenya
Korea
Latin America
Madagascar
Malaysia
Mali
Mexico
Middle East
Mozambique
Namibia
Netherlands
New Zealand
Nicaragua
Nigeria
North America
Panama
Paraguay
Peru
Philippines
Portugal
Saudi Arabia
Sierra Leone
South America

Spain
Sri Lanka
Sudan
Sweden
Switzerland
Tanzania
Thailand

Tunisia
U.S.S.R.
Uganda
Zimbabwe

Regional Files

United States Ag Policy
Alabama
Alaska
Arizona
Arkansas
California (2)
Colorado
Connecticut
Delaware
Florida
Hawaii
Illinois
Indiana
Iowa
Kansas
Kentucky
Louisiana
Maine
Maryland
Massachusetts
Michigan
Minnesota (4)
Missouri
Montana
Nebraska
New Hampshire
New Jersey
New Mexico
New York
North Carolina
North Dakota
Northeast U.S.
Ohio

Oregon
Pennsylvania
Rhode Island
South Carolina
South Dakota
Texas
Vermont
Virginia
Washington D.C.
Wisconsin

Organization File List

AAFRC
Action for Corporate Accountability
Action for Development
AFGRO - Agency to Facilitate the Growth of Rural Organizations
Africa Focus Project
Africa News
Africa Tree Center
African Development Foundation
AgAccess
Agency for International Development
AGRECOL
Agricenter International Memphis, Tennessee
Agriculture, Man & Energy
Agricultural Missions
Agriland Concepts, Inc.
AgriSystems International
Alfalfa's Market
Alley Cat Allies
Alliance for Our Common Future
Alliance for Philippine Concerns
Alliance Pour L'Enfance
Alternative Agriculture News
Alternative Payannes
Alternativodlarnas Riksforbund
American Academy for Advancement of Science (AAAS)
American Agricultural Economics Association
American Association for Corporate Contributions
American Community Garden Association
American Express Co.
American Farm Business
American Farmland Trust
American National Standards Institute
American Refugee Committee
American Society of Composer, Author and Publishers (ASCAP)
American Soil Products
Americans for Safe Food
Amoco Corp.
Animal Aid
Animal Management = Tierhaltung
Animal Rights Coalition
Animal Welfare Institution
Anthroposophic Press

Appen Features
Appropriate Technology Project
APREMA - SC
Aprovecho Institute
ARABLE - The Association for Regional Agriculture
ARIDL - The Institute for Applied Research in Drylands
Arizona Toxic Information
Arrowhead Mills
The Arts of Peace
Asian Vegetable Research & Development Center
Association of Nigerian Co-op Exporters Ltd.
ARC - Association for Retarded Citizens
Associazione Suolo E Salute
ATTRA
Audubon Society
Augsburg College
AURI
Aveda Corp.

BCS Metro
Bear Creek Farm Nursery
Bear Tribe
Bearitos
Believe
Ben & Jerry's
Beyond Beef
Bharatiya Agro-Industries Foundation
Big Mountain Support Group
"BINHI"
Bio-dynamic Farming and Gardening Association, Inc.
Biofarm
Bio-Integral Resource Center
Biolure
Biological Diversity Task Force
Biological Farming News
Board on Agriculture National Research Council
Both Ends
Bremer Foundation, Otto
Brundtland Bulletin
Brot fur die Welt
Buzzworm

California Action Network
California Agrarian Action Project

California Association of Family Farmers
California Certified Organic Farmers (CCOF)
University of California - Berkeley
University of California - Davis
University of California - Santa Cruz
Calvert Group File
The Campaign for Child Survival
Campus Outreach Opportunity League
Canadian Consulate General
Canterbury Downs
Canton Mills, Inc. (Fertilizer by Shur-Gro)
Cape Cod Coalition for Safe Food
Carleton College
Carpenter St. Croix Valley Nature Center
Catholic Charities
Catholic Institution of International Relations
CAST - Council for Agricultural Science and Technology
Catalyst
CADI (Center for Alternative Development Initiatives)
Center for Applied Studies
Center for Our Common Future
Center for Global Service and Education
Center for Rural Affairs
Center for Science in the Public Interest
Central America Resource Center (CARC)
Centre for Environment, Technology, and Development. Malaysia
Centre Europe - Tiers Monde (CETIM)
Centro de Estudios Urbanos y Regionales (CEUR)
Centro Ecologico La Pacifica
CERES
Cheese Rustlers
Chevron Corp.
Childnet
CIDOB - Centre D'Informacio I Documentacio
CINAB - Comite Interprof. National de L' Agriculture Biologique
CIRED - Centre International de Recherche sur L'Environment et le   Development
Citizens Clearinghouse for Hazardous Wastes (CCHW)
GLADES - Consorcio Latino Americanosobre Agroecologia Desarrollo
Clean Green Packing
Clean Water Action
Clinton Street Quarterly
CU Environmental Center (CO University)
Coffman Memorial Union
Coleman Natural Meats

_____Colorado
Cattlemen's Association Comite Jean
Pain
Committee on Agri. Sustainability for Developing Countries
Committee for Sustainable Agriculture
Community Administrative Services
Community Ecology
Community Environmental Council
Community Supported Agriculture
CONCERN, Inc.
Consultants for Development Programme
Consultative Group on International Agricultural Research
Consumers Union
Coolidge Center
Co-op America
Cooperating Fund Drive
Cooperation Enterprises
Coordinating Commission on Toxics and Drugs
Coordination in Development (CODEL)
CoPIRG U of No Colorado
Cordillera People's Alliance
Cornell University
The Cornucopia Project (Rodale)
Council on Economic Priorities
Council on International and Public Affairs
Culture and Agriculture
Cultural Survival Inc.
CUSO

Demeter Institute
Dentsche Gesellschaft Fur Technische (GTZ)
Developing Countries Radio Network
The Dolphin Project
Drought Policy Review Task Force
Drumghigha - Ecological Farm & Stud
E Magazine
Earth Day
Earth First!
Earth Regeneration Society, Inc.
Earth Rescue Corps
Earth - Save Foundation
Earthkeeping
Earthwork Magazine
East Wind Community
ECHO

Ecohorizons
Ecological Agriculture Projects
Econet
Ecoropa
Eden Foods
Eleventh Commandment Fellowship
Elm Farm Research Centre
Emergency Fund Service, Inc.
Emissary Foundation International
ENDA
The Entropy and Agriculture Project
Environment and Energy Study Institute
Environment Liaison Centre
Environmental Defense Fund
Environmental Health Coalition
E-LAW Environmental Law Alliance
Environmental Policy Institute
Environmental Protection Agency (EPA)
Equal Exchange
Esbenshade Orchards
Ethical Investments, Inc.
Evans Marketing Group Inc. (EMGI)
Exp. in International Living
Exxon

FACT (Food Animals Concern Trust)
Fair Trade Campaign
Family Resource Center
Fareshare Food Program
Family Farm Defense Fund
Family Farm Organizing Resource Center
Farallones Institute
Farm Family Support Project
Farm & Food Society
Farm Journal Publishing
Farm Sanctuary
Farm Verified Organic
Farmers Assistance Board
Farmers and Consumers Cooperating Together (FACCT)
Farmworker Justice Fund
Fate of the Earth
Federation of Egalitarian Communities
Federation of Reconstructionist Congregations & Havurot
Federation of Southern Cooperatives/Land Assistance Fund

Fertimax
Film in the Twin Cities
Finca La Esperanzita Organic Farm in Nicaragua
First Nations Financial Project
Five - Continent Peace Initiative
Food and Agriculture Organization (FAO°UN)
Food and Water
The Food Circle
Food First
Food Gardens Unlimited
Food Research and Action Center (FRAC)
Foothills Peace Center
Forest Research Center
FORGE (Financing Ozarks Rural Growth and Economy)
Foundation for Advancement in Science & Education
Foundation for Ecological Agriculture
Foundation on Economic Trends
Four Winds Farm Supply
Free Trade Campaign
French Meadow Bakery
Freshwater Society
Friends Network
Friends of the Earth
Friends of the Trees
Frosty Hollow Nursery
Fruition
The Fund for Animals, Inc.
Fundraising Management Services
FUTURE

GAIA Services
GAIA Earth Alive
Gale Research
Gami Sera Sevana
GASP
GDEAC - Gestion de Ecosistemas Asociacion Civil
Georgia, U of

German Marshall Fund
Gerson Therapy
Geyser
Ghost Ranch   See also: O.J. Lougheed)
Gildeal Resource Center
Global Education Associates

Global Exchange
Global Meeting on Environment and Development
Global Tomorrow Coalition
Global Volunteers
Globescope
Glynhynad Organic Farmers
Golden Hills
Good Business
Government Accountability Project
Green Gulch Farm
Green Revolution
Green River Tools
Green Seal
Greenleaf Product Company
Green Peace
Greens
Grey Panthers
Group Against Spraying Pesticides (GASP)
Group de Recherche et D'Echanges Technologiques (GRET)
Groupe Experimental Pluridisciplinaire (GEP)
Growing Tree Gardens
Grupo Social
GTZ
Habitat Centre Eco-Centre
Hamaker-Weaver Publishers
HAP - NICA
Harvest of the Heart
Health Options
Healthmed
Healthy Harvest
Hellenist America Institute: A Patrick A. O'Dougherty Institute
Herringer Brights Research Institute
Hibernian Research Company Ltd.
Hidden Springs Nursery
Highlander Resource and Education Center
Hippocrates Health Institute
Holden Farms
Housing and Urban Development (HUD)
Hubert Humphrey Institute of Public Affairs
Human Ecology Action League (HEAL)
Humane Farming Association, The
Humane Society of the US (HSUS)
Hunger Action Coalition
Hunger Project

Hungry for Profit
Hydrogen "Conversion" Project: A Patrick O'Dougherty Project
INADES Formation
Indian Summer Festival Incorporated

Infant Formula Action Coalition (INFACT)
INFORM Center
Information Center for Low External Input and Sustainable   Agriculture (ILEIA)
Institut Tunisien De Technologie Appropriee
Institute for Agriculture and Trade Policy
Institute for the Arts of Democracy
Institute for Biological Husbandry
Institute of Cultural Affairs, The
Institute for Food and Development Policy - Food First
Institute for a Future
Institute for Horticulture and Education of the Hemp Plant
Institute of Noetic Sciences
Institute for Policy Change
Institute for Social Ecology
Institute for Sustainable Agriculture
Interfaith Action Economic Justice
International Assoc. for Exchange of Students for Technical   Experience
(IAESTE)
International Center for Agricultural Research in the Dry Areas
International Centre for Conservation Education
International Christian Youth Exchange
International Coalition for Development Action
International Council for Research in Agroforestry (ICRAF)
International Crop Science Congress
International Crops Research Institute for the Semi-arid Tropics
International Development Research Centre
International Ecological Agriculture Network
International Exposition of Rural Development (IERD)

International Federation of Organic Agriculture Movements (IFOAM)
International Institute for Environment and Development
International Institute for Bau-Biologie & Ecology, Inc.
International Genetic Resources Programme (IGRIP)
International Monetary Fund
International Nutrition and Genetics Corporation (INAG)
International Organization of Consumers Unions
International Tree Crops Institute
Iowa State University
Iowa State University Humane Society
IPIAT

Irish Organic Growers Association
!TAB
Izaak Walton League

Kaw Valley Spring
Kerr Center for Sustainable Agriculture
Kisan World
KMOM Radio
Kokokahi Hunger Mission

Ladakh Institute
Lake County Defenders
Land Institute
Land O'Lakes

Land Stewardship Project
League for Ecological Democracy
League of Ecological Earth to Human Development
League of Rural Voters
League of Women Voters, The
Learning Alliance
Leopold Center for Sustainable Agriculture
Lincoln Filene Center
Little Bear Trading Co.
Living Tao Foundation
Living Tree Center
Living Wall Gardens
L'Olam: Committee on Ecology & Judaism
Lone Star Ridge Institute
Long Branch Environmental Education Center
Lundberg Farms
Lynn Canyon Ecology Centre

Macalester College
Madison Audubon Society
Maine Organic Farmers and Growers (MOFGA)
Management Assistance Program
Manor House Agricultural Centre
Maritime Permaculture Institute, The
MCC
Meadowcreek Project
Meals for Millions
Medico Friend Circle
Mercantile Development Inc. (MDI)

Merriam Hill Center
Metro State University
Mex-American Fruit Co.
Michael Fields Agricultural Institute
Michigan State University
Midwest Abundance
Midwest Pesticide Action Network
Ministere de l' Agriculture / Quebec
Minneapolis Development Education Conference
Minneapolis Trade Office
Minnesota Artists Against Hunger
Minnesota Awareness Project
Minnesota Coalition for the Homeless
Minnesota Council of Non-Profits
Minnesota Council for Social Studies
Minnesota Department of Agriculture (MDA)
Minnesota DNR {Dept. of Natural Resources)
Minnesota Environmental Education Board (MEEB)
Minnesota EQB
Minnesota Food Association
Minnesota Food Bank
Minnesota Herbicide Coalition
Minnesota Humane Society
Minnesota Institute for Sustainable Agriculture

Minnesota Interfaith Ecology Coalition
Minnesota International Center/World Affairs Center
Minnesota International Health Volunteers
Minnesota League of Conservation Voters
Minnesota Network for Animal Concerns (MINNAC)
Minnesota Organic Certification Program
Minnesota Peace and Justice Coalition (MPJC)
Minnesota Plant, Food, and Chemicals Association
Minnesota Pollution Control Agency (MPCA)
Minnesota Project
Minnesota Public Interest Research Group {MPIRG)
Minnesota Reviews
Minnesota Social Investment Forum (MSIF)
Minnesota State Horticultural Society
Minnesota State Planning Agency
Minnesota, University of
Minnesota Women's Press
Minnesotans for Safe Food
Mississippi Market

Mississippi River Revival
Missouri Botanical Garden
MOA-Mokichi Okada Association
Monsanto Co.
Mother Earth News
Mothers and Others Against Pesticides
Mountain School, The

National Agricultural Chemicals Association (NACA)
National Agricultural Library
National Agricultural Association - Australia
National Audubon Society
National Catholic Rural Life Conference
National Center for Appropriate Technology (NCAT)
NCAT Board of Directors
National Committee for Responsive Philanthropy
National Conference of State Legislators
National Conference on Nonviolence
National Council for International Health (NCIH)
National Farmers Union
National Geographic Society
National Network to Prevent Birth Defects
National Nutritional Foods Association
National Research Council
National Wildlife Federation
Natural Foods Associates
Natural Foods Merchandiser
Natural Organic Farmers Association (NOFA)
National Products Expo West
Natural Resources Defense Council
Nature Conservancy, The
Nebraska Public Interest Campaign
Nepal Community Support Group (NECOS)
New Alchemy Institute
New Covenant Peace & Justice Center
New England Environment Conference
New England Small Farm Institute
New England Farm
New Hope Communications
New Jersey Animal Rights Alliance

New Life Farm
New Mexico Environmental Department
New York Open Center
Nitrogen Fixing Tree Association

North American Bioregional Conference (NABC)
North American Farm Alliance
North American Students of Cooperation (NASCO)
North American Vegetarian Society
North Country Co-op
North Dakota Natural Farmers Association
Northern Plains Sustainable Agriculture Society
Northwest Coalition for Alternatives to Pesticides (NCAP)
NOVIB (Netherlands Organization for Int'l Cooperation in Development)
Nutrition Action
NYPIRG (New York PIRG)
Oak Manor Milling
O'Dougherty Foundation to Refound World Culture
Office of Technology Assessment
Ohio Ecological Food and Farm Association (OEFFA)
Organic Cafe Mexico
Organic Crop Improvement Association (OCIA)
Organic Farms, Inc.
Organic Food Alliance (OFA)
Organic Food Business News
Organic Foods Production Association of North America (OFPANA)
Organic Grapes Into Wine Alliance
Organic Growers Association
Organic Growers and Buyers Association
Organic Harvest
Organic Lawncare
Organic Market News and Information Service (OMNIS)
Organic Network
Our Common Future
Oxfam
Ozark Regional Land Trust
Partners in the Environment
Peace Development Fund
Peace and Environment Project
Peacenet
Peasant Energy Institute for Symbiosis
People for Ethical Treatment of Animals (PETA)
People, Food and Land Foundation
Pepin Heights
Perception and Management of Pests and Pesticides
Permaculture Institute of North America
Pesticide Action Network International (PAN)
Pesticide Action Network: Dirty Dozen Campaign

82

Pesticide Action Network (old)
Pesticide Education and Action Project (PEAP)
Pesticide Trust, The
Philanthropy Project
Philippine Peasant Movement
Ponderosa Village
Positive Village
Positive Thinkers
Practical Farmers of Iowa
Preservation of Agricultural Lands Society
Progressive Agri-Systems
Project Environment Foundation
PRONAT
Pronatura
Public Citizen
Public Voice for Food and Health Policy

Oualitatsinstitut Naturkost
Quebec - Labrador Foundation

Rachel Carson Council
Rainbow Gatherings
Rainbow Natural Foods Distributing, Ltd.
Rainforest Action Network
Rajneeshpuram
Ramapo College
Rangpur Dinajpur Rehabilitation Services
Rare Breeds International
Recycling the Urban Forest
Results
Reuter Laboratories
Rhone-Poulenc
Ringer Corp.
Rocky Mountain Institute
Rodale Institute
Rural Advancement Fund
Rural Coalition
Rural Education Center
Rural Enterprise Institute
Rural Science and Technology Institute
Rural Virginia, Inc.
Rylands
Sacred Earth Network
S.A.F.E

Sahabat Alam Malaysia (SAM)
Saint Joan of Arc
Saint Martins Table
Saint Paul Food Resources Project
Saint Paul Growers Association, Inc.
Saint Paul Pioneer Press Dispatch - Editorial Board
San Francisco League of Urban Gardeners (S.L.U.G.)
SATIS
Save the Children
School of Living
E.F. Schumacher Society SHARE
Seeds of Change
Seed Savers Exchange
Self-help
Self-Reliance Center
Sargent Shriver Award
Seventh Generation Fund
Seward Co-op
Shalom Center, The
Shalom Seed Sanctuary
SIAC (Coastal Environments Information Service)
Sierra Club
6-S
Ski to End Hunger
Small Farm Resources Project
Small Farm Viability Project
Society for International Development
Soil Association
Soil Association of South Australia
Soil Conservation Society of America
Solviva
Southern Exposure
Southland Farmer's Market Association
Spring **Hill** Center
Star Route Farm
Steering Committee for Sustainable Agriculture
Stichting Alternatief Warenonderzoek
Stichting Ekomerk Controle (SEC)
Soney Field Farm
Stonybrook Millstone Watershed
Sunrise Ranch
Sustainable Agriculture Association of Alberta
Sustainable Agriculture Network (SAN)
Sustainable Agriculture Newsletter

Sustainable Agriculture Student Organization
Sustainable Agriculture Working Group
Sustainable Farming Association of Minnesota
Sustainable Seattle

Talavaya Center
Tele 7
Third World Institute (TWI)
Third World Network
Tilth
Together--Foundation for Global Unity
Town Forum
Tranet
Trans-Species Unlimited
Trees for Life
Trilateral Commission
Turtle Island
Twin Cities Committee for the Liberation of Southern Africa

UNAG
Unidad Tecnica Ecuatoriana Del Plan AWA (UTEPA)
Union Carbide
Union of Concerned Scientists
United Farm Workers of America (AFL-CIO)
United Nations
United Nations Centre for Human Settlements (UNCHS)
United Nations Development Fund for Women (UNIFEM)
United Nations Environmental Program (UNEP)
United Nations Non-Governmental Liaison Service
United States Department of Agriculture Research (USDA)
United States Farm News
University YMCA
Urban Corps
Urban Farm Institute
Urban Foundation
Urban Resources Systems

Utne Reader, The

Vegetarian Times
Vermont Public Interest Group (VPIRG)
Vida Sana
Volunteers in Technical Assistance (VITA)

Wadebridge Ecological Centre

Wageningen Agricultural University
Wau Ecology Institute
Wedge Co-op
Wells Communication, Inc.
Wellspring
Western Washington Toxics Coalition
Westminster Presbyterian Church
White Earthland Recovery Project
White Oak Farm
Wholefood
Wilder Foundation and Forest
Wilderness Society
Wildflowers
Windstar Foundation
Winrock International
Wisconsin Rural Development Center
Wisconsin, University of -- Center of Co-ops
Witness for Peace
Womens' International League for Peace and Freedom
Wooden Shoe Gardens
Working Assets Funding Service
Working Group on Community Right-to-Know
World Bank
World Commission on Environment and Development
World Constitution and Parliament Association
World Development Movement
World Food Assembly
World Food Council
World Health Organization (WHO)

World Hunger Alleviation Through Response to Famine (WHARF)
World Hunger Year
World Resources Institute
World Society for the Protection of Animals
World Sustainable Agriculture Association
World Wildlife Fund
Worldwatch Institute
Wysong Medical Corporation

Yale University

International Alliance for Sustainable Agriculture
Resource Library
Subject Files

Subject File Name	Cross-reference File name(s)

Acid rain
AG trade
Agricultural occupational & environmental\

                health
Agroecology
Air pollution
Alachor
Alar
Aldicarb
Alfalfa
Algeny
Allelopathy
Alternative crops
Alternative economics
Aluminum pollution
Amaranth
Ames, Bruce controversy/articles
Animals
Animal right/liberation
        subfile: feminists for animals
Antibiotics
Apples
Apprenticeships
Appropriate technology
Aquacides (Aquatic Weed)
Aquaculture
Arbor Day
Aspartane
Atmospheric deposition of pesticides
        Project Aries / Toxicology of pesticides
Atrazine
AWARE
Bacillus thuringiensis
Bacteria
Bamboo
Bananas

Beans/field peas

87

Crop diversity   Education
Crop rotation
Curriculum - Sustainable Agriculture
Dacthal
Dairy
Dairy farming
Dairy   wastes
DDT
Debt - Third World crisis
Declarations of opposition to "neg. risk"
Deep Ecology
Deer
Deforestation
Delaney Clause
Democratic Party Platform
Desertification
Development (assistance, critiques, programs)
Development education
Diapers
Diarrhea
Dichlorros
Dicofol
Dietary Supplement Health & Environment Act   Nutrition
Dioxin
"Dirty Dozen"
Disease control
Draft animals
Drins
Drought
Ducks and geese
Dursban
Earth sheltered construction
Earth Summit   UN Earth Summit
Earthworms
Eco-feminism
Ecology
Economics

Edible landscaping
Education - "standard" ag
Education - sus. ag. grade 1-12
Education - sus. ag.
Employment
Endangered species
Endosulfan

89

Energy and agriculture
Entertainers
Environment (2)
Environmental illness — Chemical sensitivity
Enzymes
Erosion
Ethanol
Ethylene Dibromide (EDB)
Eubiotica Italian farming
Eucalyptus
Extension
Fababeans
Factory farming
Farm Aid (subfile: farm bill, 1990)

FIFRA
Fish
Fleas
Flies
Food additives

Farm crisis (subfile: farm conservation, Nat'l Family Farms
Farm family (subfile: small farms)
Farm safety
Farm tools
Farmworkers
Fast food
Feedlots
Fellowships
Fertilizers
Field crops

F

ood co-ops   Crisis)
Food industry
Food irradiation
Food for Peace
Food safety
Food storage
Food and values
Forest insects
Forestry
Fruit and vegetables
Fruit fly
Gardening (subfile: organic)
Garlic
GATT
Geese

Biotechnology

Greenhouse effect

Global warming

on Profit Organization Info.

Toxics
Nutrition, Diet. Supp. Act

Toxics, Pesticide files

91

Humane sustainable agriculture
Hunger   (subfile: famine)
Hydroponics
Immune system
Incinerator
Industry
Insects
Insect control
Integrated Pest Management
International Conference Nutrition
Internship   Education
Irradiation
Irrigation
Justice
Kesterson
Labeling
Ladybugs
Lakeweed control   Aquacide
Land Reclamation
Landscaping
Law
Lawn care
Lawn diseases
Lawn pests
Lawn weeds
Lead
Lead arsenate
Leucaena
Lices
Life cycle
Lifestyles
Lindane
LISA (low input sus ag)
Livestock
Livestock pests
Lupins
Macrobiotics
Maize
Malathion
Mangos
Manure
Manure fertilization
Maple production
Meat

Mechanical control
Media
Mediterranean fruit fly
Methyl Bromide
MN Sustainable Development Initiative
Mosquito control
MSG (monosodium glutamate)
Moths and caterpillars
Mulching
Multiple Chemical Sensitivity
Mushrooms
NAS alternative agriculture
National Academy of Science 1993
        Pesticides and Children
National Organic Standards Board
National Public Radio
Native American issues
Natural Food
Nature farming
NEEM
Negligible risk
Nematodes
Nestle boycott
Networking
New publications 1993
Nitrates
No-till
Noise pollution
Non Profit Organization Info.
North American Free Trade Agreement
Nosema locustate
Nuclear war
Nuclear waste
Nutrasweet
Nutrition
Oceans
Ogallala Aquifer
Oil spills
Oral rehydration therapy
Organic farming economics
Organic farming USA
Organic food health studies
Organic meat
Organic products

Environmental Illness

Ridge Till
Grants

_____Organi
c standards/certification
Oriental vegetables
Ornamental
Ozone depletion   Global warming,Greenhouse effect
Packaging
Paraquat
Perennials
Permaculture
Pest control
Pest control - biological
Pest control - household
Pesticides - general
Pesticides - in air
Pesticides - and environment
Pesticides - export
Pesticide monitoring and inspection
Pesticide - non chemical
Pesticide - ordinance
Pesticide regulation
Pesticide Reform Legislation - 1994
Pesticide residues (on food)
Pesticides - Spanish
Pesticide - Spanish
Pesticide Use Reduction Strategies
Pesticides - water
Pesticides - workers
Pets
Philanthropy
Pigs
PIK program
Pineapple
Polyculture
Population
Potatoes
Poultry farming
Poverty
Predator control
Pyrethrum
Rabbits
Radon
Rainforest
Raspberries
Rats & mice
Recycling

Recycling energy
Regenerative Agriculture
Rice
Ridge tillage   No Till
Right to Know
Roses
Rotation
Rotational grazing
Schools
Seaweeds
Seeds (subfile: germplasm)
Shallow Ecology
Sheep
Slash and burn
Social justice
Soil conservation
Soils
Solar energy
Solar greenhouse
Soybeans
Soybean byproducts
Spanish
Species diversity/loss
Spiders
Spirulina algae
Squash family
Statistics = speech
Steger, Will
Strawberries
Sugar
Sunflowers
Sustainability
Sustainable Agriculture Curriculum
Sustainable Agriculture - Introductions and Philosophy
Sweet Potatoes
Swine
Tillage
Tobacco
Tomatoes
Toxics   Pesticide files
Transportation
Travel opportunities
Trees and shrubs
Tropical agriculture

Tropical crops
24D
UN Earth Summit (Rio, 1992)(4)   Earth Summit
Urban Gardening
Valdez Principles
Veal
Vegetables
Vegetarian
Vegetation
Vetiver
Grass
Visioning
Vitamin A+ sieve
Vivisection
Volunteering
Waste management products & services
Waste treatment - household
Wastewater treatment
Water
Water pollution
Weaver, Dennis
Weed control
Weeds
Weevils   Non-profit Organization Info.
Wetlands
Wheats
Whistleblower
Whiteflies
Wildlife preservation
Wind breaks
Wine   Aquacide
Wireworms
Women - and development, status, etc.
        (subfile: women and agriculture)
World Food Day
Yams

BIBLIOGRAPHY

Agee, Betty. American Studies Program, University of Minnesota

  American Council of Learned Societies. editors. Dictionary of Scientific Biography. New York: Charles Scribners Sons, Publishers, 1981.

Austrian Studies Newsletter. Center for Austrian Studies, University of Minnesota, Winter 1996, Volume 8, Number 1.

Bernardin, Cardinal Joseph. Archdiocese of Chicago.

Bible. Second Book of Moses, Exodus Chapter 2, Verse 22.

Birch, Charles. Los Angeles Times. unknown date.

Borowski, E. J. & Borwein, J.M. editors. HarperCollins Dictionary of Mathematics. New York: HarperPerennial, 1991.

Britain, Doyle. Department of Chemistry, University of Minnesota.

Brunell, Kent. Hennepin County Medical Center, Minneapolis.

Cicconardi, S.P., Jannelli, E., Spazzafumo, G. International Journal of Hydrogen Energy, Vol. 18, No. 11, Nov. 1993.

Conant, Brother Gregory, OSB. St. Benedict's Abbey. Still River, Massachusetts.

Duffy, Joseph. patrick in his own words. Dublin: Veritas Publications, 1985.

Editorial Adive of the Faculties of the University of Chicago. Encyclopedia Britannica, Inc. 15th. Edition. Chicago: Encyclopedia Britannica, Inc. 15th. Edition, 1990. Vol. 5. Micropedia.

Fadiman, Clifton. The Lifetime Reading Plan. New York: The World Publishing Company, 1960.

Finn, Richard. poet. St. Paul, Minnesota.

Freeman, Father Hilary. former professor of logic, St. Catherine's College, St. Paul, Minnesota. St. Thomas Aquinas Priory. River Forest, Illinois.

Ghandi, Mahatma.

Global Ratification and Elections Network. 1480 Hoyt St., Suite 31 Lakewood, Colorado.

Grayson, Martin. Executive Editor. Kirk-OthmerEncyclopedia of Chemical Technology. Third Edition. New York: John Wiley & Sons, 1981.

Guralnik, David B. editor. Webster's New World Dictionary of the American Language. New World: Simon and Schuster, 1980.

Hampl, Patricia. Virgin Time. (New York: Farrar, Straus and Giroux, 1992).

Harris, John. psychologist.

Hazwinkel, M. editor. Encyclopedia of Mathematics. Dordrecht. Netherlands: Kluwer Academic Publishing, 1988.

Holm, Jeane. Women in the Military: An Unfinished Revolution. Novato, California: Presidio Press, 1992.

International Alliance for Sustainable Agriculture. Newman Center. University of Minnesota. Minneapolis, Minnesota.

Internet reference. Topics in hydrogen use. http://www.ciesin.org/1C/wri/entrglow.htm1

Internet reference. Topic hydrogen and electricity. http://bbs.pn1.gov:2080/.oem/technical2/hydrogen.htm 1

Ireland, Patricia. President of NOW.

Kakutani, Michiko. Books of the Times Bookreview of Laurie Garrett's The Coming Plague: Newly Emerging Diseases in a World Out of Balance. New York: Farrar, Straus and Giroux, 1994.

Kemp, Michael. Internet Daily Mail article found in the Times by Kevin Easton

Lewis, Christopher. Ph.D. entitled, Progress and Apocalypse: Science and the End of the Modern World. University of Minnesota, 1991.

Mackenzie, James. The Keys to the Car: Electric and Hydrogen Vehicles for the 21st Century. World Resources Institute, 1994.

Manna. Newsletter of the International Alliance for Sustainable Agriculture.

Minneapolis, Newman Center. July-August, 1994, Vol. 11, No. 1.

McGee, Mark G. and David W. Wilson. Psychology: Science and Application. St. Paul. West Publishing Company, 1984.

Meier, Andreas, Ingo Uhlendorf and Dieter Meissner. Proceedings of the Intersociety Energy Conversion Engineering Conference. Vol. 4, 1994. IEEE, Piscataway, NJ, USA.

O'Dougherty, Aquinas. author's father.

O'Dougherty, John. author's uncle.

O'Dougherty, Loyola. author's uncle.

Peavey, Michael A. Fuel From Water: Energy Independence with Hydrogen Louisville, KY. Merit, Inc., 1993.

Pegg-Karlsson, Berit. The British-Scandinavian Association for Wind and Hydrogen Power. Hydrogen fuel from water Internet reference.

Petermeier, Jerry. The Wienery. 4th and Cedar, Minneapolis, Minnesota.

Poulson, Barry W. World Book Encyclopedia. Vol 1. Chicago, 1996.

Rand, Ayn. The Fountainhead. New York. Bobbs-Merrill, 1943.

Rosa, V.M. and M.B.F. Santos and E.P. Da Silva. International Journal of Hydrogen Energy. Vol. 20, No. 9, Sep. 1995.

Sessions, George. editor. Deep Ecology for the 21st. Century: Readings on the Philosophy and Practice of the New Environmentalism Boston. Shambhala, 1995.

Swimme Brian and Thomas Berry. The Universe Story. New York: HarperSanFrancisco A Division of HarperCollins Publishers, 1992.

Teresa, Mother. A Simple Path. New York. Ballantine Books, 1995.

Thompson, Neale. Hennepin County Medical Center. Minneapolis, Minnesota.

Travasio, Joseph. Minneapolis, Minnesota.

Wagner, Sister Mary Anthony. Oblates of St. Benedict. College of St. Benedict. St. Joseph, Minnesota.

Walker, Alice. The Color Purple. New York. Harcourt Brace Jovanovich.

Webster's New Universal Unabridged Dictionary. New York. Barnes & Noble Books, 1994.

Wojtyla, Karol (Pope John Paul II). Love and Responsibility. New York. Farrar, Straus and Giroux, 1981.

Water producer 33

106

Footnotes

Introduction Footnotes

1. Joseph Duffy, patrick: in his own words (Dublin: Veritas Publications, 1985), 35-36.
2. Ibid., Appendix Two, 98-99.
3. Ibid., 9-11.
4. "Of Human Bondage," is the name of an essay by Baruch Spinoza and the title of a book by Sommerset Maugham. This reference comes from Clifton Fadiman's, The Lifetime Reading Plan (New York: The World Publishing Company, 1960), 232-233.

5. George Sessions, editor, <u>Deep Ecology for the 21st. Century: Readings on the Philosophy and Practice of the New Environmentalism</u> (Boston: Shambhala, 1995), 294-296.
6. <u>Ibid</u>.
7. Jerry Petermeier gave the inhabitable to habitable definition of terra forma. He owns the Wienery on 4th and Cedar in Minneapolis, Minnesota.
8. Sessions, <u>op. cit.</u> 241.
9. Patricia Hampl, <u>Virgin Time</u> (New York: Farrar, Straus and Giroux, 1992), title.
10. M. Hazwinkel, editor, "cardinality," <u>Encyclopedia of Mathematics.</u> (Dordrecht, Netherlands: Kluwer Academic Publishing, 1988) Vol.2,24.
11. E.J. Borowski & J.M. Borwein, editors, "cardinality," <u>HarperCollins Dictionary of Mathematics</u> (New York: HarperPerennial, 1991), 67
12. "Hierarchy of cardinalities," <u>Ibid.</u>, 67.
13. Barry W. Poulson, "Kenneth Joseph Arrow," "Impossibility Theorem," <u>World Book Encyclopedia</u> (Chicago, 1996), Vol. 1, 743.
14. George Sessions, <u>op. cit.</u> ix.

15.   Ibid., xi-xiii.
16.   Second Book of Moses, Exodus Chapter 2, Verse 22.
17.   Brother Gregory Conant, OSB, St. Benedict's Abbey, 252 Still River Road, Still River Massachusetts 01461.
18.   Webster's New Universal UnAbridged Dictionary, "steward, stewardship," (New York: Barnes & Noble Books, 1994), 1395.
19.   Ibid.,26-27.
20.   Mark McGee, a psychologist friend.
21.   Mark G. McGee and David W. Wilson, Psychology: Science and Application (St. Paul: West Publishing Company, 1984), 6-7.
22.   Sessions, op. cit., ix-x.
23.   Msgr. Loyola O'Dougherty, an uncle, San Manuel, Arizona.
24.   Sessions, op. cit., x.
25.   Ibid., xi.
26.   Ibid., xiv-xviii.
27.   Ibid., xviii-xix.
28.   Ibid., xx-xxi.
29.   Ibid., 3-4.
30.   Ibid., 4-5.
31.   Richard Finn, poet, St. Paul, Minnesota.
32.   Sessions, op. cit., 4-5.
33.   Ibid., 4-7.
34.   Ibid., 8-9.
35.   Ibid., 8-9.
36.   Ibid., 9-10.
37.   Ibid., 12-15.
38.   Betty Agee, American Studies Program, University of Minnesota, mentioned the example of redwood forest destruction.
39.   Sessions, op. cit., 16-18.
40.   Cardinal Joseph Bernadin, Archdiocese of Chicago, argues for a "seamless garment" approach to the pro-life issue.
41.   Sessions, op. cit., 19-24.
42.   Richard Murray provides this definition of the Gaia Hypothesis.
43.   Sessions, op. cit., 24-25.
44.   Ayn Rand in The Fountainhead uses the term "collective soul." An American rock band adapted this concept for its title. This is where the writer came across this reference.
45.   Sessions, op. cit., 24-25.
46.   Ibid., 27-29.
47.   Ibid., 36-37.
48.   Ibid., 37-39.

49.   Ibid., 38-39.
50.   Ibid., 39-40.
51.   Ibid., 41-42.

52.   Ibid., 42-44.
53.   Ibid., 42-45.
54.   Ibid., 45-49.
55.   Ibid., 50-62.
56.   Ibid., 68.
57.   The writer and Joe Travasio, a Minneapolis, Minnesota, leftwing thinker.
58.   Sessions, op. cit., 97-100.
59.   Ibid., 93-94.
60.   Ibid.
61.   Ibid., 100-102.
62.   Ibid., 104-105.
63.   Ibid., 105-108.
64.   Ibid., 108-110.
65.   Ibid., 113-114.
66.   Ibid., 113-115.
67.   Ibid., 117.
68.   Ibid., 122.
69.   Ibid., 122-124.
70.   Ibid., 122-126.
71.   Ibid., 131-133.
72.   Ibid., 133-136.
73.   Ibid., 137.
74.   Ibid., 138.
75.   John N. Harris, a psychologist.

76.   Sessions, op. cit., 139.
77.   Ibid., 139-140.
78.   Ibid., 141-147.

79. Ibid., 151-154.
80. Ibid., 154-155.
81. Ibid., 157-158.
82. Dr. Kent Brunell, Hennepin County Medical Center, Minneapolis, Minnesota.
83. Sessions, op. cit., 159-170.
84. Ibid., 170-173.
85. Ibid., 188.
86. Ibid., 187-193.
87. Ibid., 241.
88. Ibid., 205-245.
89. Ibid., 246-248.
90. Ibid., 249-258.
91. Ibid., 257-261.

Footnotes to <u>A Simple Path</u> and Saint Benedict

1. Mother Teresa, <u>A Simple Path</u> (New York: Ballantine Books, 1995), xi-xxxvIll.
2. Sister Mary Anthony Wagner, editor, "SAINT BENEDICT," Oblates of St. Benedict, College of St. Benedict, St. Joseph, Minnesota.

Footnotes to the New Age Movement and the Green Party Movement

1. Sessions, <u>op. cit.</u>, 266-267.
2. Patricia Ireland, in a private conversation.
3. Sessions, <u>op. cit.</u>267-268.
4. <u>Ibid.</u>, 272-278.
5. <u>Ibid.</u>, 293.
6. <u>Ibid.</u>, 293-294.
7. <u>Ibid.</u>, 294-298.
8. <u>Ibid.</u>, 323-324.
9. Aquinas O'Dougherty, my father.
10. Neale Thompson, Hennepin County Medical Center, Minneapolis, Minnesota.
11. Sessions, <u>op. cit.</u>324-329.
12. <u>Ibid.</u>, 328-329.
13. Aquinas O'Dougherty, my father.
14. Anonymous.
15. Sessions, <u>op. cit.</u>364-366.
16. <u>Ibid.</u>, 367.
17. <u>Ibid..381.</u>
18. <u>Ibid.</u>, lost reference.
19. John O'Dougherty, an uncle uses the phrase cultural garbage in reference to American pop culture.
20. Sessions, <u>op. cit.</u> 413-416.
21. Father Hilary Freeman, St. Thomas Aquinas Priory, River Forest, Illinois.
22. Aquinas O'Dougherty, my father.
23. John N. Harris, psychologist.
24. John O'Dougherty, an uncle.
25. Christopher Lewis, a Ph.D., entitled, Progress and Apocalypse: Science and the End of the Modern World. University of Minnesota, 1991.
26. Mother Teresa of India.

Footnotes to Apocalyptic Visions

1. Michiko Kakutani, Books of the Times Bookreview of Laurie Garrett's, <u>The Coming Plague: Newly Emerging Diseases in a World Out of Balance</u> (New York: Farrar, Straus and Giroux, 1994), deleted page number.
2. <u>Ibid</u>.

3.   Charles Birch, "As humans send Earth toward extinction," Los Angeles Times. unknown date.

Footnotes to Catholic Personalism: An Interpretative Approach to the Green Revolution

1.   David B. Guralnik, editor, "personalism," <u>Webster's New World Dictionary of the American Language</u> (New York: Simon & Schuster, 1980), 1062.
2.   John O'Dougherty, my uncle.
3.   Conant, O.S.B., <u>op. cit.</u> unpublished letter entitled, "Our Almighty God," 1-2.
4.   <u>Ibid</u>.
5.   Conant, O.S.B., in a private conversation suggested this Biblical reference.
6.   Guralnik, <u>op. cit.</u> "language," 792.
7.   <u>Ibid.</u>, "mathematics," 875.
8.   Father Hilary Freeman maintains that mathematicians cannot agree on how to define their field.
9.   Gurlanik, <u>op. cit.</u> "methodology," 895.

Footnotes to Female Intuition and the Intuitionist School of Mathematics

1.   Guralnik, <u>op. cit.</u> "intuition," 740.
2.   <u>Ibid</u>.
3.   <u>Ibid.</u>, "intuitionism," 740.
4.   Brother Gregory Conant, O.S.B., <u>op. cit.</u> in a private conversation.
5.   American Council of Learned Societies, editors, "Brouwer, Luitzen Egbertus Jan," <u>Dictionary of Scientific Biography</u> (New York: Charles Scribners Sons, Publishers, 1981), Vol. 11.,512.
6.   <u>Ibid</u>.
7.   Pope John Paul 11, Karol Wojtyla, <u>Love and Responsibility</u> (New York: Farrar, Straus, Giroux, 1981), 23.

6.   Webster's New Universal Unabridged Dictionary, "hydrogen," (New York: Barnes & Noble Books, 1994), 696.
7.   Ibid., "sea," 1285.
8.   Ibid., "ocean," 996.
9.   Ibid., "Neptune," 959.
10.   Internet reference Berit Pegg-Karlsson, the British-Scandinavian Association for Wind and Hydrogen Power, supported by the Pure Energy Trust.
11.   Ibid.
12.   Ibid.
13.   Internet, English reference adapted from an Daily Mail article by Michael Kemp and found in the Times by Kevin Easton.
14.   Ibid.
15.   Brother Gregory Conant, OSB, op. cit.
16.   Editorial Advice of the Faculties of the University of Chicago, "Hindenburg," Encyclopedia Britannica Inc. 15th. Edition (Chicago: Encyclopedia Britannica, Inc.,1990), Micropedia, Vol. 5, p. 933.
17.   Michael A. Peavey, Fuel From Water: Energy Independence with Hydrogen (Louisville, KY: Merit, Inc., 1993), 12.
18.   Ibid.
19.   See Patrick A. O'Dougherty, Personalism and Mathematics as Women's Personifestoes (Minneapolis: The Hellenist America Institute, 1995), 14.
20.   Peavey, op. cit.. 13-14.
21.   Ibid..15-19.
22.   Ibid., 21.
23.   Ibid., 29-49.
24.   Ibid.
25.   Ibid., 52-53.
26.   Ibid., 53-57.
27.   Ibid., 57-65.
28.   Ibid., 65-67.
29.   Ibid., 68.
30.   Ibid., 75-76.
31.   O'Dougherty, op. cit.. 14.
32.   Peavey, op. cit.. 76-77.
33.   Ibid., 77.
34.   Ibid., 78.
35.   Ibid., 78.
36.   Ibid., 78.
37.   Ibid., 79.
38.   Ibid., 79.
39.   Ibid.

40. <u>Ibid.</u>, 79.
41. <u>Ibid.</u>, 84.

42. Ibid., 84-87.
43. Ibid., 87.
44. Ibid., 91-92.
45. Ibid., 92.
46. Ibid., 94.
47. Ibid., 94-101.
48. Ibid., 105.
49. Ibid., 108.
50. Ibid., 112-114.
51. Ibid..116-118.
52. O'Dougherty, op. cit., 15.
53. Peavey, op. cit., 121.
54. Ibid., 121.
55. Ibid., 124.
56. Ibid., 134.
57. Ibid., 140.
58. Ibid., 141.
59. Ibid., 142-144.
60. Ibid., 144.
61. Ibid..146.
62. Ibid., 149.
63. Ibid..149-150.
64. Ibid., 154.
65. Ibid..154-158.
66. Ibid., 161.
67. Ibid., 161-164.
68. Ibid., 164.
69. Ibid., 183.
70. Ibid., 185.
71. Ibid., 185-187.
72. Ibid., 188.
73. Ibid., 188.
74. Ibid., 188-190.
75. Ibid., 190.
76. Ibid., 190-199.
77. Ibid., 200.
78. Ibid., 200.
79. Ibid., 200-201.
80. Ibid., 201.
81. Ibid., 201-202.
82. Ibid., 202-204.
83. Ibid., 204-206.
84. Ibid., 214.

85.   Ibid., 216-220.

86. Ibid., 223.
87. Ibid., 224-225.
88. Ibid., 229-234.

Footnotes to the Neptune Project: Water, Salt and Sun

1. V.M. Rosa, M.B.F. Santos, E.P. Da Silva, International Journal of Hydrogen Energy, Vol. 20, No. 9, Sep. 1995, 697-700.
2. Professor Doyle Britain, Department of Chemistry, University of Minnesota.
3. Martin Grayson, Executive Editor.Kirk-OthmerEncyclopedia of Chemical Technology. Third Edition, "membrane technology," Table 3. "Energy Consumption for Desalting Brackish Water (10,000 ppm), By Various Technologies,"(New York: John Wiley & Sons, 1981), Vol. 15, 114.

Footnotes to Hydrogen Versus Electric Vehicles

1. James Mackenzie, The Keys to the Car: Electric and Hydrogen Vehicles for the 21st Century (World Resources Institute, 1994), vi-vii.
2. Ibid., 1-2.
3. Ibid..3-11.
4. Ibid..11-16.
5. Ibid., 17.
6. Ibid., 17-19.
7. Ibid., 21-24.
8. Ibid., 24.
9. Ibid., 25.
10. Ibid., 25-26.
11. Ibid., 30.
12. Ibid., 31.
13. Ibid., 31-33.
14. Ibid., 31-36.
15. Ibid., 34.
16. Ibid., 35-36.
17. Ibid..35-36.
18. Ibid., 36-39.
19. Ibid., 38-39.
20. Ibid., 39-40.
21. Ibid., 42.
22. O'Dougherty, op. cit.. 14-21.
23. Mackenzie, op. cit., 43-52.
24. Ibid., 52.
25. Ibid., 52-55.

26. Ibid., 62-63.
27.
Ibid., 71.
28. Ibid., 75-79.
29. Ibid., 81-82.
30. Ibid., 82-99.
31. S.P. Cicconardi, E. Jannelli, G. Spazzafumo, International Journal of Hydrogen Energy, Vol. 18, No. 11, Nov. 1993, 933-940.
32. Andreas Meier, Ingo Uhlendorf, Dieter Meissner, Proceedings of the Intersociety Energy Conversion Engineering Conference, Vol. 4, 1994. IEEE, Piscataway, NJ, USA, 1697-1702.
33. Ibid., 61-63.
34. Internet reference A series of topics on hydrogen use. http://www.ciesin.org/IC/wri/entrglow.html
35. Internet reference on hydrogen and electric use htt://bbs.pnl.gov:2080/.oem/technica12/hydrogen.html

Footnotes to the Common Good

1. Abraham Joshua Heschel's, The Sabbath, proclaims Judaism is "religion of time" and the Sabbath sanctifies time. However, hydrogen projects should sanctify both time and space.
2. John Paul 11, op. cit., 117-118.
3. Ibid.
4. David B. Guralnik, op. cit., "solidarity," 1355.
5. John Paul II, op. cit., 147.
6. T.S. Eliot, lost reference.
7. John Paul II, op. cit., 118-119.
8. Brother Gregory Conant, O.S.B., op. cit.
9. Aquinas O'Dougherty, my father.
10. Global Ratification and Elections Network "Partial List of Organizations included in the Global Ratification and Elections Network" GREN, 1480 Hoyt St., Suite 31, Lakewood, Colorado, 80215-4755.
11. Austrian Studies Newsletter, Center for Austrian Studies, University of Minnesota, Winter 1996, Volume 8, Number 1, p. 9.

Footnotes to a Definition of Right or Rights from Father Hilary Freeman

1. Father Hilary Freeman, op. cit., unpublished letter, dated November 4, 1995.
2. Ibid.
3. Brother Gregory Conant, OSB.
4. CHARTER OF THE RIGHTS OF CATHOLICS IN THE CHURCH, Association for the Rights of Catholics in the Church, P.O. Box 912, Delran, N.J., 08075.

5.   Ibid.
6.   Ibid.
7.   Cardinal Joseph Bernardin, Archdiocese of Chicago.
8.   Jeane Holm, Women in the Military: An Unfinished Revolution (Novato, California: Presidio Press, 1992).
9.   Mahatma Ghandi, this quote was found on a St. Thomas College, St. Paul, Minnesota, bulletin board.
10. Jerry Petermeier gave the inhabitable to habitable definition of terra forma. He owns the Wienery on 4th and Cedar in Minneapolis, Minnesota.

Footnotes to the International Alliance for Sustainable Agriculture: Newman Center: A Case Study--A New Cycle

1.   . Manna, Newsletter of the International Alliance for Sustainable Agriculture (Minneapolis: Newman Center) July-August, 1994, Vol. 11, No.1, 8.
2   Ibid., 2.
3.   Ibid.
4.   Ibid., 3.

Footnotes to the Ecozoic Age

1.   Brian Swimme & Thomas Berry, The Universe Story (New York: HarperSanFrancisco A Division of HarperCollins Publishers, 1992).
2.   Ibid., 242-243.
3.   Ibid., 243.
4.   Ibid., 243-249.
5.   Ibid., 253-258.

Footnotes to the IASA Purple Data Base

1. International Alliance for Sustainable Agriculture Newman Center at the University of Minnesota, 1701 University Avenue SE, Minneapolis, MN 55414 USA.
2. Alice Walker, <u>The Color Purple</u> (New York: Harcourt Brace Jovanovich, 1982).

116

www.ingramcontent.com/pod-product-compliance
Lightning Source LLC
Chambersburg PA
CBHW032002170526
45157CB00002B/503